Use R!

Use R!

This series of inexpensive and focused books on R will publish shorter books aimed at practitioners. Books can discuss the use of R in a particular subject area (e.g., epidemiology, econometrics, psychometrics) or as it relates to statistical topics (e.g., missing data, longitudinal data). In most cases, books will combine LaTeX and R so that the code for figures and tables can be put on a website. Authors should assume a background as supplied by Dalgaard's Introductory Statistics with R or other introductory books so that each book does not repeat basic material.

More information about this series at http://www.springer.com/series/6991

Eric D. Kolaczyk · Gábor Csárdi

Statistical Analysis of Network Data with R

Second Edition

 Springer

Eric D. Kolaczyk
Department Mathematics and Statistics
Boston University
Boston, MA, USA

Gábor Csárdi
RStudio
Boston, MA, USA

ISSN 2197-5736   ISSN 2197-5744   (electronic)
Use R!
ISBN 978-3-030-44128-9   ISBN 978-3-030-44129-6   (eBook)
https://doi.org/10.1007/978-3-030-44129-6

1st edition: © Springer Science+Business Media New York 2014
2nd edition: © Springer Nature Switzerland AG 2020
Chapter 11 is adapted in part with permission from Chapter 4 of Eric D. Kolaczyk, Topics at the Frontier of Statistics and Network Analysis: (Re)Visiting the Foundations, SemStat Elements (Cambridge: Cambridge University Press, 2017), doi:10.1017/9781108290159.

This Springer imprint is published by the registered company Springer Nature Switzerland AG
The registered company address is: Gewerbestrasse 11, 6330 Cham, Switzerland

Pour Josée, sans qui ce livre n'aurait pas vu le jour—E.D.K.

Z-nak, a gyémántért és az aranyért—G.CS.

Preface

Networks and network analysis are arguably one of the largest growth areas of the early twenty-first century in the quantitative sciences. Despite roots in social network analysis going back to the 1930s, and roots in graph theory going back centuries, the phenomenal rise and popularity of the modern field of `network science', as it is sometimes called, is something that could not have been predicted 20 years ago. Networks have permeated everyday life, far beyond the realm of research and methodology, through now-familiar realities such as the Internet, social networks, viral marketing, and more.

Measurement and data analysis are integral components of network research. As a result, there is a critical need for all sorts of statistics for network analysis, both common and sophisticated, ranging from applications, to methodology, to theory. As with other areas of statistics, there are both descriptive and inferential statistical techniques available, aimed at addressing a host of network-related tasks, including basic visualization and characterization of network structure; sampling, modeling, and inference of network topology; and modeling and prediction of network-indexed processes, both static and dynamic.

Software for performing most such network-related analyses is now available in various languages and environments, across different platforms. Not surprisingly, the R community has been particularly active in the development of software for doing statistical analysis of network data. As of this writing there are already dozens of contributed R packages devoted to some aspect of network analysis. Together, these packages address tasks ranging from standard manipulation, visualization, and characterization of network data (e.g., **igraph**, **network**, and **sna**), to modeling of networks (e.g., **igraph**, **eigenmodel**, **ergm**, and **blockmodels**), to network topology inference (e.g., **glasso** and **huge**). In addition, there is a great deal of analysis that can be done using tools and functions from the R base package.

In this book we aim to provide an easily accessible introduction to the statistical analysis of network data, by way of the R programming language. As a result, this book is not, on the one hand, a detailed manual for using the various R packages encountered herein, nor, on the other hand, does it provide exhaustive coverage of the conceptual and technical foundations of the topic area. Rather, we have attempted to strike a balance between the two and, in addition, to do so using a (hopefully!) optimal level of brevity. Accordingly, we envision the book being used, for example, by (i) statisticians looking to begin engaging in the statistical analysis of network data, whether at a research level or in conjunction with a new collaboration, and hoping to use R as a natural segue, (ii) researchers from other similarly quantitative fields (e.g., computer science, statistical physics, and economics) working in the area of complex networks, who seek to get up to speed relatively quickly on how to do statistical analyses (both familiar and unfamiliar) of network data in R, and (iii) practitioners in applied areas wishing to get a foothold on how to do a specific type of analysis relevant to a particular application of interest.

More generally, the book has been written at a level aimed at graduate students and researchers in quantitative disciplines engaged in the statistical analysis of network data, although advanced undergraduates already comfortable with R should find much of the book fairly accessible as well. Therefore, we anticipate the book being of interest to readers in statistics, of course, but also in areas such as computational biology, computer science and machine learning, economics, neuroscience, quantitative finance, signal processing, statistical physics, and the quantitative social sciences.

For the second edition of this book, there are three significant changes. First, following a package-wide overhaul of the nomenclature used in **igraph** a few years ago, all of the many calls to **igraph** functions throughout this book have been updated accordingly. Second, we have added a new chapter on the topic of networked experiments, an area in which there has been an explosion of recent activity, with relevance from the health sciences to politics to marketing. Lastly, mirroring the substantial amount of research and development on the topic of stochastic block models over the past 5 years, we have updated our treatment to incorporate the **blockmodels** package.

There are a number of people we wish to thank, whose help at various stages of development and writing is greatly appreciated. Thanks again go to the editorial team at Springer for their enthusiasm in encouraging us to take on this project originally and to pursue the current revision. Thanks go as well to the various students in the course Statistical Analysis of Network Data (MA703) at Boston University in the Fall semesters of 2013, 2015, and 2019 for their comments and feedback. Special thanks for this edition are due to Will Dean and Jiawei Li, who spent the better part of a summer going through every code line in the book for functionality, nomenclature, and such. We are again grateful as well to Christophe Ambroise, Alain Barrat, Mark Coates, Suchi Gopal, Emmanuel Lazega, and Petra Staufer for kindly making available their data.

More broadly, we would like to express our appreciation in general for the countless hours of effort invested by the developers of the many R packages that we have made use of throughout the pages of this book. Without their work, the breadth and scope of our own here would be significantly reduced. And we would like to thank the many people who have made use of the first edition of the book and sent feedback, whether in the form of enthusiastic comments or flags for new software glitches that arose. Finally, yet again, we wish to express our deepest gratitude to our respective families for their love, patience, and support throughout the revision of this book.

All code and data used in this book have been made available in the R package **sand**, distributed through the CRAN archive.

Boston, MA, USA   Eric D. Kolaczyk
London, England   Gábor Csárdi
December 2019

Contents

Chapter 1
Introduction

1.1 Why Networks?

The oft-repeated statement that "we live in a connected world" perhaps best captures, in its simplicity, why networks have come to hold such interest in recent years. From on-line social networks like Facebook to the World Wide Web and the Internet itself, we are surrounded by examples of ways in which we interact with each other. Similarly, we are connected as well at the level of various human institutions (e.g., governments), processes (e.g., economies), and infrastructures (e.g., the global airline network). And, of course, humans are surely not unique in being members of various complex, inter-connected systems. Looking at the natural world around us, we see a wealth of examples of such systems, from entire eco-systems, to biological food webs, to collections of inter-acting genes or communicating neurons.

The image of a network—that is, essentially, something resembling a *net*—is a natural one to use to capture the notion of elements in a system and their inter-connectedness. Note, however, that the term 'network' seems to be used in a variety of ways, at various levels of formality. The *Oxford English Dictionary,* for example, defines the word *network* in its most general form simply as "a collection of inter-connected things." On the other hand, frequently 'network' is used inter-changeably with the term 'graph' since, for mathematical purposes, networks are most commonly represented in a formal manner using graphs of various kinds. In an effort to empha-size the distinction between the general concept and its mathematical formalization, in this book we will use the term 'network' in its most general sense above, and—at the risk of the impression of a slight redundancy—we will sometimes for emphasis refer to a graph representing such a network as a 'network graph.'

The seeds of network-based analysis in the sciences, particularly its mathematical foundation of graph theory, are often placed in the 1735 solution of Euler to the now famous Königsberg bridge problem, in which he proved that it was impossible to walk the seven bridges of that city in such a way as to traverse each only once. Since then, particularly since the mid-1800s onward, these seeds have grown in a number of key areas. For example, in mathematics the formal underpinnings were

© Springer Nature Switzerland AG 2020

E. D. Kolaczyk and G. Csárdi, *Statistical Analysis of Network Data with R*, Use R!, https://doi.org/10.1007/978-3-030-44129-6_1

systematically laid, with König [5] cited as the first key architect. The theory of electrical circuits has always had a substantial network component, going back to work of Kirchoff, and similarly the study of molecular structure in chemistry, going back to Cayley. As the fields of operations research and computer science grew during the mid-1900s, networks were incorporated in a major fashion in problems involving transportation, allocation, and the like. And similarly during that time period, a small subset of sociologists, taking a particularly quantitative view towards the topic of social structure, began developing the use of networks in characterizing interactions within social groups.

More recently—starting, perhaps, in the early to mid-1990s—there was an explosion of interest in networks and network-based approaches to modeling and analysis of complex systems. Much of the impetus for this growth derived from work by researchers in two particular areas of science: statistical physics and computer science. To the former can be attributed a seminal role in encouraging what has now become a pervasive emphasis across the sciences on understanding how the interacting behaviors of constituent parts of a whole system lead to collective behavior and systems-level properties or outcomes. Indeed the term *complex system* was coined by statistical physicists, and a network-based perspective has become central to the analysis of complex systems. To the latter can be attributed much of the theory and methodology for conceptualizing, storing, manipulating, and doing computations with networks and related data, particularly in ways that enable efficient handling of the often massive quantities of such data. Moreover, information networks (e.g, the World Wide Web) and related social media applications (e.g., Twitter), the development of which computer scientists have played a key role, are examples of some of the most studied of complex systems (arguably reflecting our continued fascination with studying ourselves!).

More broadly, a network-based perspective has now been found to be useful in the study of complex systems across a diverse range of application areas. These areas include computational biology (e.g., studying systems of interacting genes, proteins, chemical compounds, or organisms), engineering (e.g., establishing how best to design and deploy a network of sensing devices), finance (e.g., studying the interplay among, say, the world's central banks as part of the global economy), marketing (e.g., assessing the extent to which product adoption can be induced as a type of 'contagion'), neuroscience (e.g., exploring patterns of voltage dynamics in the brain associated with epileptic seizures), political science (e.g., studying how voting preferences in a group evolve in the face of various internal and external forces), and public health (e.g., studying the spread of infectious disease in a population, and how best to control that spread).

In general, two important contributing factors to the phenomenal growth of interest in networks over the past two decades are (i) a tendency towards a systems-level perspective in the sciences, away from the reductionism that characterized much of the previous century, and (ii) an accompanying facility for high-throughput data collection, storage, and management. The quintessential example is perhaps that of the changes in biology in the modern era, during which the complete mapping of the human genome, a triumph of computational biology in and of itself, in turn paved the way for fields like systems biology to be pursued aggressively, wherein a detailed understanding is sought of how the components of the human body, at the molecular level and higher, work together.

Ultimately, the study of systems such as those just described is accompanied by measurement and, accordingly, the need for statistical analysis. The focus of this book is on how to use tools in R to do statistical analysis of *network data*. More specifically, we aim to present tools for performing what are arguably a core set of analyses of measurements that are either of or from a system conceptualized as a network.

1.2 Types of Network Analysis

Network data are collected daily in a host of different areas. Each area, naturally, has its own unique questions and problems under study. Nevertheless, from a statistical perspective there is a methodological foundation, composed of tasks and tools that are each common to some non-trivial subset of research areas involved with network science. Furthermore, it is possible—and indeed quite useful—to categorize many of the various tasks faced in the analysis of network data across different domains according to a statistical taxonomy. It is along the lines of such a taxonomy that this book is organized, progressing from descriptive methods to methods of modeling and inference, with the latter conveniently separated into two sub-areas, corresponding to the modeling and inference of networks themselves versus processes on networks. We illustrate with some examples.

1.2.1 Visualizing and Characterizing Networks

Descriptive analysis of data (i.e., as opposed to statistical modeling and inference) typically is one of the first topics encountered in a standard introductory course in statistics. Similarly, the visualization and numerical characterization of a network usually is one of the first steps in network analysis. Indeed, in practice, descriptive analyses arguably constitute the majority of network analyses published.

Consider the network in Fig. 1.1. Shown is a visualization[1] of part of a dataset on collaborative working relationships among members of a New England law firm, collected by Lazega [6]. These data were collected for the purpose of studying cooperation among social actors in an organization, through the exchange of various types of resources among them. The organization observed was a law firm, consisting of over 70 lawyers (roughly half partners and the other half associates) in three offices located in three different cities. Relational data reflecting resource exchange were collected, and additional attribute information was recorded for each lawyer, includ-

---

[1] The R code for generating this visualization is provided in Chap. 3.

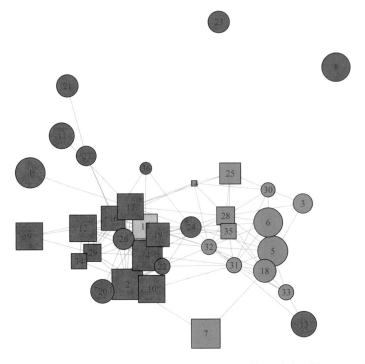

Fig. 1.1 Visualization of Lazega's network of collaborative working relationships among lawyers. *Vertices* represent partners and are labeled according to their seniority (with 1 being most senior). Vertex colors (i.e., *red*, *blue*, and *yellow*) indicate three different office locations, while vertex shape corresponds to the type of practice [i.e., litigation (*circle*) and corporate (*square*)]. Vertex area is proportional to number of years with the law firm. *Edges* indicate collaboration between partners. There are three female partners (i.e., those with seniority labels 27, 29, and 34); the rest are male

ing type of practice, gender, and seniority. See Lazega and Pattison [7] for additional details.

The visual summary in Fig. 1.1 manages to combine a number of important aspects of these data in one diagram. A graph is used to represent the network, with vertices corresponding to lawyers, and edges, to collaboration between pairs of lawyers. In addition, differences in vertex color, shape, size, and label are used to indicate office location, type of practice, years with the law firm, and seniority. However, this is far from the only manner in which to visualize these data. Deciding how best to do so is itself both an art and science. Furthermore, visualization is aided here by the fact that the network of lawyers is so small. Suppose instead our interest lay in a large on-line social network, such as the Facebook network of people that have 'friended' each other. With over 2.4 billion active users reported as of this writing, it is impossible to display this network in an analogous manner, with every individual user and their friendships evident. A similar statement often can be made in biology, for example, such as for networks of proteins and their affinity for binding or of genes and their regulatory relationships. As a result, the visualization of large networks is a separate challenge of its own. We will look at tools for network visualization in Chap. 3.

Characterization of network data through numerical summaries is another important aspect of descriptive analysis for networks. However, unlike in an introductory statistics course, where the types of summaries encountered typically are just simple measures of center (e.g., mean, median, etc.) and dispersion (e.g., standard deviation, range, etc.) for a set of real numbers, summary measures for networks necessarily seek to capture characteristics of a graph. For example, for the lawyer data in Fig. 1.1, it is natural to ask to what extent two lawyers that both work with a third lawyer are likely to work with each other as well. This notion corresponds to the social network concept of *transitivity* and can be captured numerically through an enumeration of the proportion of vertex triples that form triangles (i.e., all three vertex pairs are connected by edges), typically summarized in a so-called clustering coefficient. In this case, the relevant characterization is an explicit summary of network structure. On the other hand, given the two types of lawyers represented in these data (i.e., corporate and litigation), it is also natural to ask to what extent lawyers of each type collaborate with those of same versus different types. This notion corresponds to another social network concept—that of *assortativity*—and can be quantified by a type of correlation statistic (the so-called assortativity coefficient), in which labels of connected pairs of vertices are compared. In this case, the focus is on an attribute associated with network vertices (i.e., lawyer practice) and the network structure plays a comparatively more implicit role.

There are a host of network summaries available, with more still being defined. A variety of such measures are covered in Chap. 4. These range from characterization of properties associated with individual vertices or edges to properties of subgraphs to properties of the graph as a whole. The trick to employing network characterizations in a useful fashion generally is in matching the question(s) of interest in an underlying complex system with an appropriate summary measure(s) of the corresponding network.

1.2.2 Network Modeling and Inference

Beyond asking what an observed network looks like and characterizing its structure, at a more fundamental level we may be interested in understanding how it may have arisen. That is, we can conceive of the network as having resulted from some underlying processes associated with the complex system of interest to us and ask what are the essential aspects of these processes. In addition, the actual manner in which the network was obtained, i.e., the corresponding measurement and construction process, may well be important to take into consideration. Such concerns provide the impetus for network modeling and associated tools of statistical inference.

Network modeling has received much attention. Broadly speaking, there are two classes of network models: mathematical and statistical. By 'mathematical models' we mean models specified through (typically) simple probabilistic rules for generating a network, where often the rules are defined in an attempt to capture a particular mechanism or principle (e.g., 'the rich get richer'). In contrast, by 'statistical models' we mean models (often probabilistic as well) specified at least in part with the intention that they be fit to observed data—allowing, for example, the evaluation of the explanatory power of certain variables on edge formation in the network—and that the fit be effected and assessed using formal principles of statistical inference. While certainly there is some overlap between these two classes of models, the relevant literatures nevertheless are largely distinct.

The simplest example of a mathematical network model is one in which edges are assigned randomly to pairs of vertices based on the result of a collection of independent and identically distributed coin tosses—one toss for each vertex pair. Corresponding to a variant of the famous Erdős–Rényi formulation of a random graph, this model has been studied extensively since the 1960s. Its strength is in the fact not only that its properties are so well understood (e.g., in terms of how, for example, cohesive structure emerges as a function of the probability of an edge) but also in the role it plays as a standard against which to compare other, more complicated models.

From the statistical perspective, however, such mathematical models generally are too simple to be a good match to real network data. Nevertheless, they are not only useful in allowing for formal insight to be gained into how specific mechanisms of edge formation may affect network structure, but they also are used commonly in defining null classes of networks against which to assess the 'significance' of structural characteristics found in an observed network. For example, in Fig. 1.1, the clustering coefficient for the network of lawyer collaboration turns out to be just slightly less than 0.40. Is this value large or not? It is difficult to say without a frame of reference. One way of imposing such a frame of reference is to define a simple but comparable class of networks (e.g., all graphs with the same numbers of vertices and edges) and look at the distribution of the clustering coefficient across all such graphs. In the spirit of a permutation test and similar data re-use methods, examining where our observed value of 0.40 falls within this distribution gives us a sense of how unusual it is in comparison to values for networks across the specified class. This approach has been used not only for examining clustering in social networks but also, for example, in identifying recurring sub-graph patterns (aka 'motifs') in gene regulatory networks and functional connectivity networks in neuroscience. We will explore the use of mathematical network models for such purposes in Chap. 5.

There are a variety of statistical network models that have been offered in the literature, many of which parallel well-known model classes in classical statistics. For example, exponential random graph models are analogous to generalized linear models, being based on an exponential family form. Similarly, latent network models, in specifying that edges may arise at least in part from an unmeasured (and possibly unknown) variable(s), directly parallel the use of latent variables in hierarchical statistical modeling. Finally, stochastic block models may be viewed as a form of mixture model. Nevertheless, importantly, the specification of such models and their fitting typically are decidedly less standard, given the usually high-dimensional and dependent nature of the data. Such models have been used in a variety of settings, from the modeling of collaborations like that in Lazega's lawyer network to the prediction of protein pairs likely to have an affinity for physically binding. We will look at examples of the use of each of these classes of models in Chap. 6.

1.2.3 Network Processes

As objects for representing interactions among elements of a complex system, network graphs are frequently the primary focus of network analysis. On the other hand, in many contexts it is actually some quantity (or attribute) associated with each of the elements in the system that ultimately is of most interest. Nevertheless, in such settings it often is not unreasonable to expect that this quantity be influenced in an important manner by the interactions among the elements, and hence the network graph may still be relevant for modeling and analysis. More formally, we can picture a stochastic process as 'living' on the network and indexed by the vertices in the network. A variety of questions regarding such processes can be interpreted as problems of prediction of either static or dynamic network processes.

Returning to Fig. 1.1, for example, suppose that we do not know the practice of a particular lawyer. It seems plausible to suspect that lawyers collaborate more frequently with other lawyers in the same legal practice. If so, knowledge of collaboration may be useful in predicting practice. That is, for our lawyer of unknown practice, we may be able to predict that practice with some accuracy if we know (i) the vertices that are neighbors of that lawyer in the network graph, and (ii) the practice of those neighbors.

While in fact this information is known for all lawyers in our data set, in other contexts we generally are not so fortunate. For example, in on-line social networks like Facebook, where users can choose privacy settings of various levels of severity, it can be of interest (e.g., for marketing purposes) to predict user attributes (e.g., perhaps indicative of consumer preferences) based on that of 'friends'. Similarly, in biology, traditionally the functional role of proteins (e.g., in the context of communication within the cell or in controlling cell metabolism) has been established through labor-intensive experimental techniques. Because proteins that work together to effect certain functions often have a higher affinity for physically binding to each other, the clusters of highly connected proteins in protein–protein interaction networks can be exploited to make computational predictions of protein function, by propagating information on proteins of known function to their neighbors of unknown function in the graph. We will encounter methods for making such predictions of static network processes in Chap. 8.

Ultimately, many (most?) of the systems studied from a network-based perspective are intrinsically dynamic in nature. Not surprisingly, therefore, many processes defined on networks are more accurately thought of as dynamic, rather than static, processes. Consider, for example, the context of public health and disease control. Understanding the spread of a disease (e.g., the H1N1 flu virus) through a population can be modeled as the diffusion of a binary dynamic process (indicating infected or not infected) through a network graph, in which vertices represent individuals, and edges, contact between individuals. Mathematical modeling (using both deterministic and stochastic models) arguably is still the primary tool for modeling such processes, but network-based statistical models gradually are seeing increased use, particularly as better and more extensive data on contact networks becomes available. Another setting in which analogous ideas and models are used—and wherein it is possible to collected substantially more data—is in modeling the adoption of new products among consumers (e.g., toward predicting the early adopters of, say, the next iPhone released). We look briefly at techniques in this still-developing area of modeling and prediction of dynamic network processes in Chap. 8 as well.

In a related direction, we will look at statistical methods for the modeling and prediction of network *flows* in Chap. 9. Referring to the movement of something—materials, people, or commodities, for example—from origin to destination, flows are a special type of dynamic process fundamental to transportation networks (e.g., airlines moving people) and communication networks (e.g., the Internet moving packets of information), among others.

Alternatively, a more recently emerged area of research involving network processes is that of *networked experiments*, wherein the goal—rather than prediction—is to account for network connectivity in assessing (and, often, exploiting!) the influence of the elements in a complex system on each other in an inter-connected system. Across the sciences—social, biological, and physical alike—there is a pervasive interest in evaluating the causal effect of treatments or interventions of various kinds. In Chap. 10, we look at the assessment of experimental treatment effects within networked systems.

Finally, in Chap. 11, we will look briefly at the area of dynamic network analysis, wherein the network, the process(es) on the network, or indeed both, are expected to be evolving in time.

1.3   Why Use R for Network Analysis?

Various tools are available for network analysis. Some of these are standalone programs, like the Windows-based Pajek tool or the java-based Gephi, while others are embedded into a programming environment and are essentially used as a programming library. Some examples of the latter are NetworkX in Python and **igraph** in R.

R has become the de facto standard of statistical research. The majority of new statistical developments are immediately available in R and no other programming languages or software packages. While network analysis is an interdisciplinary field, it is not an exception from this trend. Several R extension packages implement one or more network analysis algorithms or provide general tools for manipulating network data and implement network algorithms. R supports high quality graphics and virtually any common graphical file format. R is extensible with currently over 10,000 extension packages, and this number continues to grow exponentially.

Being a complete programming language, R offers great flexibility for network research. New network analysis algorithms can be prototyped rapidly by building on the existing network science extension packages, the most commonly used one of which is the **igraph** package. In addition to implementations of classic and recently published methods of network analysis, **igraph** provides tools to import, manipulate and visualize graphs, and can be used as a platform for new algorithms. It is currently used directly in over four hundred other R packages and will be featured often in this book.

1.4 About This Book

Our goal in writing this book is to provide an easily accessible introduction to the statistical analysis of network data, by way of the R programming language. The book has been written at a level aimed at graduate students and researchers in quantitative disciplines engaged in the statistical analysis of network data, although advanced undergraduates already comfortable with R should find the book fairly accessible as well. At present, therefore, we anticipate the book being of interest to readers in statistics, of course, but also in areas like computational biology, computer science and machine learning, digital humanities, economics, neuroscience, quantitative finance, signal processing, statistical physics, and the quantitative social sciences.

The material in this book is organized to flow from descriptive statistical methods to topics centered on modeling and inference with networks, with the latter separated into two sub-areas, corresponding first to the modeling and inference of networks themselves, and then, to processes on networks. More specifically, we begin by covering tools for the manipulation of network data in Chap. 2. The visualization and characterization of networks is then addressed in Chaps. 3 and 4, respectively. Next, the topics of mathematical and statistical network modeling are investigated, in Chaps. 5 and 6, respectively. In Chap. 7 the focus is on the special case of network modeling wherein the network topology must itself be inferred. Network processes, both static and dynamic, are addressed in Chap. 8, while network flows are featured in Chap. 9. Networked experiments—wherein the network process of interest is perturbed through treatment or intervention—is the topic of Chap. 10. Finally, in Chap. 11 a brief look is provided at the extent to which the topics of these earlier chapters have been extended to the context of dynamic networks.

It is not our intent to provide a comprehensive coverage here of the conceptual and technical foundations of the statistical analysis of network data. For something more of that nature, we recommend the book by Kolaczyk [3].[2] There are also a number of excellent treatments of network analysis more generally, written from the perspective of various other fields. These include the book by Newman [8], from the perspective of statistical physics, and the book by Jackson [2], from the perspective

---

[2]Additionally, a succinct presentation of statistical topics at the frontier of network analysis may be found in the recent monograph by Kolaczyk [4].

of economics. The book by Easley and Kleinberg [1], written at an introductory undergraduate level, provides a particularly accessible treatment of networks and social behavior. Finally, the book by Wasserman and Faust [10], although less recent than these others, is still an important resource, particularly regarding basic tools for characterization and classical network modeling, from a social network perspective.

On the other hand, neither is this book a complete manual for the various R packages encountered herein. The reader will want to consult the manuals for these packages themselves for details omitted from our coverage. In addition, it should be noted that we assume a basic familiarity with R and the base package. For general background on R, particularly for those not already familiar with the software, there are any number of resources to be had, including the classic tutorial by Venebles and Smith [9] and the more recent book by Wickham and Grolemund [11].

Ultimately, we have attempted to strike a reasonable balance between the concepts and technical background, on the one hand, and software details, on the other. Accordingly, we envision the book being used, for example, by (i) statisticians looking to begin engaging in the statistical analysis of network data, whether at a research level or in conjunction with a new collaboration, and hoping to use R as a natural segue; (ii) researchers from other similarly quantitative fields (e.g., computer science, statistical physics, economics, etc.) working in the area of complex networks, who seek to get up to speed relatively quickly on how to do statistical analyses (both familiar and unfamiliar) of network data in R, and (iii) practitioners in applied areas wishing to get a foothold into how to do a specific type of analysis relevant to a particular application of interest.

1.5 About the R Code

The R code in this book requires a number of data sets and R packages. We have collected the data sets into a standalone R package called **sand** (i.e., for 'statistical analysis of network data'). If the reader aims to run the code while reading the book, she needs to install this package from CRAN.[3] Once installed, the package can then be used to also install all other R packages required in the code chunks of the book:

```
#1.1 1 > install.packages("sand")
      2 > library(sand)
      3 > install_sand_packages()
```

To avoid the need for constant typing while reading the book, we have extracted all R code and placed it online, at http://github.com/kolaczyk/sand. We suggest that the reader open this web page to simply copy-and-paste the code chunks into an R session. All code is also available in the **sand** package. In the package manual are details on how to run the code:

---

[3]The Comprehensive R Archive Network (CRAN) is a worldwide network of ftp and web servers from which R code and related materials may be downloaded. see http://cran.us.r-project.org/.

```
#1.2 1 > ?sand
```

Note that code chunks are not independent, in that they often rely upon the results of previous chunks, within the same chapter. On the other hand, code in separate chapters is meant to be run separately—in fact, restarting R at the beginning of each chapter is the best way to ensure a clean state to start from. For example, to run the code in Chap. 3, it is *not* required (nor recommended) to first run all code in this chapter and Chap. 2, but it *is* prudent to run any given code block in Chap. 3 only after having run all other relevant code blocks earlier in the chapter.

In writing this book, we have sought to strike an optimal balance between reproducibility and exposition. It should be possible, for example, for the reader to exactly reproduce all numerical results presented herein (i.e., summary statistics, estimates, etc.). In order to ensure this capability, we have set random seeds where necessary (e.g., prior to running methods relying on Markov chain Monte Carlo). The reader adapting our code for his own purposes will likely, of course, want to use his own choice of seeds. In contrast, however, we have compromised when it comes to visualizations. In particular, while the necessary code has been supplied to repeat all visualizations shown in this book, in most cases we have not attempted to wrap the main code with all necessary supplemental code (e.g., random seeds, margin settings, space between subplots, etc.) required for exact reproducibility, which would have come at what we felt was an unacceptable cost of reduction in readability and general aesthetics.

Given the fact that the R environment in general, and the R packages used in this book more specifically, can be expected to continue to evolve over time, we anticipate that we will need to update the online version of the code (and the version in the **sand** package) in the future, to make it work with newer versions of R and with updated packages. Therefore, if the reader finds that the code here in the book appears to fail, the online version should be most current and working.

On a related note, as can be seen above, the code chunks in the book are numbered, by chapter and within each chapter. This convention was adopted mainly to ensure ease in referencing code chunks from the website and vice versa. Even if we update the code on the web site, our intention is that the chunk numbers will stay the same.

In addition to having the most current version of the code chunks, the web page also includes an issue tracker, for reporting mistakes in the text or the code, and for general discussion.

References

1. D. Easley and J. Kleinberg, *Networks, Crowds, and Markets*. Cambridge Univ Press, 2010.
2. M. Jackson, *Social and Economic Networks*. Princeton Univ Pr, 2008.
3. E. Kolaczyk, *Statistical Analysis of Network Data: Methods and Models*. Springer Verlag, 2009.
4. E. D. Kolaczyk, *Topics at the Frontier of Statistics and Network Analysis: (Re) Visiting the Foundations*. Cambridge University Press, 2017.

5. D. König, *Theorie der Endlichen und Unendlichen Graphen*. American Mathematical Society, 1950.
6. E. Lazega, *The Collegial Phenomenon: The Social Mechanisms of Cooperation Among Peers in a Corporate Law Partnership*. Oxford: Oxford University Press, 2001.
7. E. Lazega and P. Pattison, "Multiplexity, generalized exchange and cooperation in organizations: a case study," *Social Networks*, vol. 21, no. 1, pp. 67–90, 1999.
8. M. Newman, *Networks: an Introduction*. Oxford University Press, Inc., 2010.
9. W. Venables and D. Smith, "An introduction to R," *Network Theory Ltd*, 2009.
10. S. Wasserman and K. Faust, *Social Network Analysis: Methods and Applications*. New York: Cambridge University Press, 1994.
11. H. Wickham and G. Grolemund, *R for Data Science: Import, Tidy, Transform, Visualize, and Model Data*. "O'Reilly Media, Inc.", 2016.

Chapter 2
Manipulating Network Data

2.1 Introduction

We have seen that the term 'network,' broadly speaking, refers to a collection of elements and their inter-relations. The mathematical concept of a graph lends precision to this notion. We will introduce the basic elements of graphs—both undirected and directed—in Sect. 2.2 and discuss how to generate network graphs, both 'by hand' and from network data of various forms.

As a representation of a complex system, a graph alone (i.e., as merely a collection of vertices and edges) is often insufficient. Rather, there may be additional information important to the application at hand, in the form of variables that can be indexed by the vertices (e.g., gender of members of a social network) or the edges (e.g., average time required to traverse a link in a transportation network). Alternatively, at a coarser level of granularity, it may be convenient to associate vertices or edges with groups (e.g., all proteins in a protein–protein interaction network that are involved with a certain type of signaling event in a cell). Indeed, we can imagine potentially equipping vertices and edges with several variables of interest. Doing so corresponds to the notion of decorating a network graph, which is discussed in Sect. 2.3.

Finally, in using graphs to represent network data, a certain level of familiarity with basic graph theoretic concepts, as well as an ability to assess certain basic properties of graphs, is essential. We therefore devote Sect. 2.4 to a brief overview of such concepts and properties, including a quick look at a handful of important special classes of graphs.

For creating, decorating, and assessing basic properties of network graphs, **igraph** is particularly useful.[1] A library and R package for network analysis, **igraph** contains a set of data types and functions for (relatively!) straightforward implementation and

---

[1] Alternatively, there is within the **network** and **sna** packages, found in the **statnet** suite, a similarly rich set of tools for the manipulation and characterization of network graphs. These packages share nontrivial overlap with **igraph**.

© Springer Nature Switzerland AG 2020

E. D. Kolaczyk and G. Csárdi, *Statistical Analysis of Network Data with R*, Use R!,
https://doi.org/10.1007/978-3-030-44129-6_2

rapid prototyping of graph algorithms, and allows for the fast handling of large graphs (e.g., on the order of millions of vertices and edges). As such, its use will figure heavily in this and the following two chapters (i.e., where the emphasis is on descriptive methods). The fact that **igraph** was developed as a research tool and that its focus originally was to be able to handle large graphs efficiently, means that its learning curve used to be somewhat steep. Recent versions do not necessarily flatten the learning curve, but are nevertheless friendlier to the user, once she has mastered the basics.

2.2 Creating Network Graphs

2.2.1 Undirected and Directed Graphs

Formally, a *graph* $G = (V, E)$ is a mathematical structure consisting of a set V of *vertices* (also commonly called *nodes*) and a set E of *edges* (also commonly called *links*), where elements of E are unordered pairs $\{u, v\}$ of distinct vertices $u, v \in V$. The number of vertices $N_v = |V|$ and the number of edges $N_e = |E|$ are sometimes called the *order* and *size* of the graph G, respectively. Often, and without loss of generality,[2] we will label the vertices simply with the integers $1, \ldots, N_v$, and the edges, analogously.

In **igraph** there is an 'igraph' class for graphs.[3] In this section, we will see a number of ways to create an object of the igraph class in R, and various ways to extract and summarize the information in that object.

For small, toy graphs, the function `graph_from_literal` can be used, specifying the edges in a symbolically literal manner. For example,

```
#2.1  1  > library(igraph)
       2  > g <- graph_from_literal(1-2, 1-3, 2-3, 2-4, 3-5, 4-5,
       3  +                         4-6, 4-7, 5-6, 6-7)
```

creates a graph object g with $N_v = 7$ vertices

```
#2.2  1  > V(g)
       2  + 7/7 vertices, named, from fac8b33:
       3  [1] 1 2 3 4 5 6 7
```

and $N_e = 10$ edges

```
#2.3  1  > E(g)
       2  + 10/10 edges from fac8b33 (vertex names):
       3  [1] 1--2 1--3 2--3 2--4 3--5 4--5 4--6 4--7 5--6 6--7
```

---

[2] Technically, a graph G is unique only up to relabellings of its vertices and edges that leave the structure unchanged. Two graphs that are equivalent in this sense are called *isomorphic*.

[3] The exact representation of 'igraph' objects is not visible for the user and is subject to change.

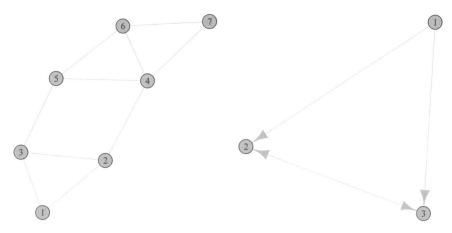

Fig. 2.1 *Left*: an undirected graph. *Right*: a directed graph

This same information, combined and in a slightly different format, is recovered easily using the function `print_all`.[4]

```
#2.4  1 > print_all(g)
      2 IGRAPH fac8b33 UN-- 7 10 --
      3 + attr: name (v/c)
      4 + edges (vertex names):
      5 1 -- 2, 3
      6 2 -- 1, 3, 4
      7 3 -- 1, 2, 5
      8 4 -- 2, 5, 6, 7
      9 5 -- 3, 4, 6
     10 6 -- 4, 5, 7
     11 7 -- 4, 6
```

A visual representation of this graph, generated simply through the command[5]

```
#2.5  1 > plot(g)
```

is shown in Fig. 2.1, on the left.

The character U seen accompanying the summary of g above indicates that our graph is *undirected*, in that there is no ordering in the vertices defining an edge. A graph G for which each edge in E has an ordering to its vertices (i.e., so that $\{u, v\}$ is distinct from $\{v, u\}$, for $u, v \in V$) is called a *directed graph* or *digraph*. Such edges are called *directed edges* or *arcs*, with the direction of an arc $\{u, v\}$ read from left to right, from the *tail* u to the *head* v. Note that digraphs may have two arcs between

---

[4]Variations on this output are obtained, alternatively, using `print` or `summary`. Note that in applying these and other similar functions to a graph, the resulting output includes an identifier (here `fac8b33`) associated uniquely with the graph by **igraph**. In fact, an abbreviation of the identifier is returned; the full identifier can be obtained using the `graph_id` function.

[5]This is the most basic visualization. We will explore the topic of visualization on its own in more depth in Chap. 3.

a pair of vertices, with the vertices playing opposite roles of head and tail for the respective arcs. In this case, the two arcs are said to be *mutual*.

Directed edges in `graph_from_literal` are indicated using a minus/plus convention. In Fig. 2.1, on the right, is shown an example of a digraph consisting of three vertices, with two directed edges and one mutual edge.

```
#2.6  1  > dg <- graph_from_literal(1-+2, 1-+3, 2++3)
       2  > plot(dg)
```

We note that in defining both of the graphs above we have used the standard convention of labeling vertices with the numbers 1 through N_v, which is also the default in **igraph**. In practice, however, we may already have natural labels, such as the names of people in a social network, or of genes in a gene regulatory network. Such labels can be used instead of the default choice by generating the graph with them explicitly.

```
#2.7  1  > dg <- graph_from_literal(Sam-+Mary, Sam-+Tom,
      2  +                          Mary++Tom)
      3  > print_all(dg)
      4  IGRAPH a21f0d9 DN-- 3 4 --
      5  + attr: name (v/c)
      6  + edges from a21f0d9 (vertex names):
      7  [1] Sam ->Mary Sam ->Tom  Mary->Tom  Tom ->Mary
```

Alternatively, vertex labels can be changed from the default after initially creating the graph, by modifying the `name` attribute of the graph object.

```
#2.8  1  > V(dg)$name <- c("Sam", "Mary", "Tom")
```

2.2.2 Representations for Graphs

Realistically, we do not usually expect to enter a graph by hand, since most networks encountered in practice have at least tens of vertices and edges, if not tens of thousands (or even millions!). Rather, information for constructing a network graph typically will be stored in a data file. At the most elementary level, there are three basic formats: adjacency lists, edge lists, and adjacency matrices.

An *adjacency list* representation of a graph G is simply an array of size N_v, ordered with respect to the ordering of the vertices in V, each element of which is a list, where the ith list contains the set of all vertices j for which there is an edge from i to j. This is the representation usually used by **igraph**, evident in the output from the function `print_all` in the examples above.

An *edge list* is a simple two-column list of all vertex pairs that are joined by an edge. In **igraph**, edge lists are implicit, for example, in returning the edge set E.

```
#2.9  1 > E(dg)
      2 + 4/4 edges from 062bf79 (vertex names):
      3 [1] Sam ->Mary Sam ->Tom  Mary->Tom  Tom ->Mary
```

The function as_edgelist returns an edge list as a two-column R matrix.

Finally, graphs can also be stored in matrix form. The $N_v \times N_v$ *adjacency matrix* for a graph $G = (V, E)$, say **A**, is defined so that

$$A_{ij} = \begin{cases} 1, & \text{if } \{i, j\} \in E, \\ 0, & \text{otherwise}. \end{cases} \qquad (2.1)$$

In words, **A** is non-zero for entries whose row-column indices (i, j) correspond to vertices in G joined by an edge, from i to j, and zero, for those that are not. The matrix **A** will be symmetric for undirected graphs.

```
#2.10  1 > as_adjacency_matrix(g)
       2 7 x 7 sparse Matrix of class "dgCMatrix"
       3   1 2 3 4 5 6 7
       4 1 . 1 1 . . . .
       5 2 1 . 1 1 . . .
       6 3 1 1 . . 1 . .
       7 4 . 1 . . 1 1 1
       8 5 . . 1 1 . 1 .
       9 6 . . . 1 1 . 1
      10 7 . . . 1 . 1 .
```

This last choice of representation is often a natural one, given that matrices are fundamental data objects in most programming and software environments and that network graphs frequently are encoded in statistical models through their adjacency matrices. However, their use with the type of large, sparse networks commonly encountered in practice can be inefficient, unless coupled with the use of sparse matrix tools.

In **igraph**, network data already loaded into R in these specific formats can be used to generate graphs using functions such as graph_from_adj_list, graph_from_edgelist, and graph_from_adjacency_matrix, respectively. For data stored in a file, the function read_graph can be used. In fact, this latter function not only supports the three formats discussed above, but also a number of other formats (e.g., such as GraphML, Pajek, etc.). Conversely, the function write_graph can be used to save graphs in various formats.

2.2.3 Operations on Graphs

A graph(s) that we are able to load into R may not be the graph that we ultimately want. Various operations on the graph(s) we have available may be necessary, including extracting part of a graph, deleting vertices, adding edges, or even combining multiple graphs.

The notion of a 'part' of a graph is captured through the concept of a subgraph. A graph $H = (V_H, E_H)$ is a *subgraph* of another graph $G = (V_G, E_G)$ if $V_H \subseteq V_G$ and $E_H \subseteq E_G$. Often we are interested in an *induced subgraph* of a graph G, i.e., a subgraph $G' = (V', E')$, where $V' \subseteq V$ is a prespecified subset of vertices and $E' \subseteq E$ is the collection of edges to be found in G among that subset of vertices. For example, consider the subgraph of g induced by the first five vertices.

```
#2.11  1  > h <- induced_subgraph(g, 1:5)
       2  > print_all(h)
       3  IGRAPH 2560ed9 UN-- 5 6 --
       4  + attr: name (v/c)
       5  + edges from 2560ed9 (vertex names):
       6  [1] 1--2 1--3 2--3 2--4 3--5 4--5
```

The inclusion or exclusion of vertices or edges in a graph $G = (V, E)$ can be conceived of as the application of addition or subtraction operators, respectively, to the sets V and E. For example, the subgraph h generated just above could also have been created from g by removing the vertices 6 and 7.

```
#2.12  1  > h <- g - vertices(c(6,7))
```

Similarly, g can be recovered from h by first adding these two vertices back in, and then, adding the appropriate edges.

```
#2.13  1  > h <- h + vertices(c(6,7))
       2  > g <- h + edges(c(4,6),c(4,7),c(5,6),c(6,7))
```

Finally, the basic set-theoretic concepts of union, disjoint union, intersection, difference, and complement all extend in a natural fashion to graphs. For example, the union of two graphs, say H_1 and H_2, is a graph G in which vertices and edges are included if and only if they are included in at least one of H_1 or H_2. For example, our toy graph g may be created through the union of the (induced) subgraph h defined above and a second appropriately defined subgraph.

```
#2.14  1  > h1 <- h
       2  > h2 <- graph_from_literal(4-6, 4-7, 5-6, 6-7)
       3  > g <- union(h1,h2)
```

2.3 Decorating Network Graphs

2.3.1 Vertex, Edge, and Graph Attributes

At the heart of a network-based representation of data from a complex system will be a graph. But frequently there are other relevant data to be had as well. From a network-centric perspective, these other data can be thought of as *attributes*, i.e., values associated with the corresponding network graph. Equipping a graph with such attributes is referred to as *decorating* the graph. Typically, the vertices or edges of a graph (or both) are decorated with attributes, although the graph as a whole may be decorated as well. In **igraph**, the elements of graph objects (i.e., particularly the vertex and edge sequences, and subsets thereof) may be equipped with attributes simply by using the ' $ ' operator.

Vertex attributes are variables indexed by vertices, and may be of discrete or continuous type. Instances of the former type include the gender of actors in a social network, the infection status of computers in an Internet network in the midst of an on-line virus (e.g., a worm), and a list of biological pathways in which a protein in a protein–protein interaction network is known to participate, while an example of the latter type is the voltage potential levels in the brain measured at electrodes in an electrocorticogram (ECoG) grid. For example, recall that the names of the three actors in our toy digraph are

```
#2.15 1 > V(dg)$name
      2 [1] "Sam"   "Mary"  "Tom"
```

Their gender is added to dg as

```
#2.16 1 > V(dg)$gender <- c("M","F","M")
```

Note that the notion of vertex attributes also may be used advantageously to equip vertices with properties during the course of an analysis, either as input to or output from calculations within R. For example, this might mean associating the color red with our vertices

```
#2.17 1 > V(g)$color <- "red"
```

to be used in plotting the graph (see Chap. 3). Or it might mean saving the values of some vertex characteristic we have computed, such as the types of vertex centrality measures to be introduced in Chap. 4.

Edge attributes similarly are values of variables indexed by adjacent vertex pairs and, as with vertex attributes, they may be of both discrete or continuous type. Examples of discrete edge attributes include whether one gene regulates another in an inhibitory or excitatory fashion, or whether two countries have a friendly or antagonistic political relationship. Continuous edge attributes, on the other hand, often represent some measure of the strength of relationship between vertex pairs. For example, we might equip each edge in a network of email exchanges (with vertices representing email addresses) by the rate at which emails were exchanged over a given period of time. Or we might define an attribute on edges between adjacent stations in a subway network (e.g., the Paris metro) to represent the average time necessary during a given hour of the day for trains to run from one to station to the next.

Often edge attributes can be thought of usefully, for the purposes of various analyses, as weights. Edge weights generally are non-negative, by convention, and often are scaled to fall between zero and one. A graph for which the edges are equipped with weights is referred to as a *weighted graph*.[6]

---

[6]More generally, a weighted graph can be defined as a pair (V, E), where V is a set of vertices, as before, but the elements in E are now non-negative numbers, with one such number for each vertex

```
#2.18 1 > is_weighted(g)
      2 [1] FALSE
      3 > wg <- g
      4 > E(wg)$weight <- runif(ecount(wg))
      5 > is_weighted(wg)
      6 [1] TRUE
```

As with vertex attributes, edge attributes may also be used to equip edges with properties to be used in calls to other R functions, such as the `plot` function.

In principle, a graph itself may be decorated with an attribute, and indeed, it is possible to equip graph objects with attributes in **igraph**. The most natural use of this feature arguably is to equip a graph with relevant background information, such as a name

```
#2.19 1 > g$name <- "Toy Graph"
```

or a seminal data source.

2.3.2 Using Data Frames

Just as network graphs typically are not entered by hand for graphs of any nontrivial magnitude, but rather are encoded in data frames and files, so too attributes tend to be similarly encoded. For example, in R, a network graph and all vertex and edge attributes can be conveniently represented using two data frames, one with vertex information, and the other, with edge information. Under this approach, the first column of the vertex data frame contains the vertex names (i.e., either the default numerical labels or symbolic), while each of the other columns contain the values of a given vertex attribute. Similarly, the first two columns of the edge data frame contain an edge list defining the graph, while each of the other columns contain the values of a given edge attribute.

Consider, for example, the lawyer data set of Lazega [5], introduced in Chap. 1. Collecting the information on collaborative working relationships, in the form of an edge list, in the data frame `elist.lazega`, and the various vertex attribute variables, in the data frame `v.attr.lazega`, they may be combined into a single graph object in **igraph** as

```
#2.20 1 > library(sand)
      2 > g.lazega <- graph_from_data_frame(elist.lazega,
      3 +                                     directed="FALSE",
      4 +                                     vertices=v.attr.lazega)
      5 > g.lazega$name <- "Lazega Lawyers"
```

---

pair. Analogously, the adjacency matrix **A** for a weighted graph is defined such that the entry A_{ij} is equal to the corresponding weight for the vertex pair i and j.

Our full set of network information on these

```
#2.21 1 > vcount(g.lazega)
      2 [1] 36
```

lawyers now consists of the

```
#2.22 1 > ecount(g.lazega)
      2 [1] 115
```

pairs that declared they work together, along with the eight vertex attributes

```
#2.23 1 > vertex_attr_names(g.lazega)
      2 [1] "name"      "Seniority" "Status"    "Gender"
      3 [5] "Office"    "Years"     "Age"       "Practice"
      4 [9] "School"
```

(in addition to the vertex name).[7]

We will see a variety of ways in the chapters that follow to characterize and model these network data and others like them.

2.4   Talking About Graphs

2.4.1   *Basic Graph Concepts*

With the adoption of a graph-based framework for representing relational data in network analysis we inherit a rich vocabulary for discussing various important concepts related to graphs. We briefly review and demonstrate some of these here, as they are necessary for doing even the most basic of network analyses.

As defined at the start of this chapter, a graph has no edges for which both ends connect to a single vertex (called *loops*) and no pairs of vertices with more than one edge between them (called *multi-edges*). An object with either of these properties is called a *multi-graph*.[8] A graph that is not a multi-graph is called a *simple* graph, and its edges are referred to as *proper* edges.

It is straightforward to determine whether or not a graph is simple. Our toy graph g is simple.

```
#2.24 1 > is_simple(g)
      2 [1] TRUE
```

---

[7]The functions vcount and ecount are aliases for the functions gorder and gsize in **igraph**, which return the order and size of the input graph, respectively.

[8]In fact, the **igraph** data model is more general than described above, and allows for multi-graphs, with multiple edges between the same pair of vertices and edges from a vertex to itself.

But duplicating the edge between vertices 2 and 3, for instance, yields a multi-graph.

```
#2.25  1 > mg <- g + edge(2,3)
       2 > print_all(mg)
       3 IGRAPH f00a980 UN-- 7 11 -- Toy Graph
       4 + attr: name (g/c), name (v/c), color (v/c)
       5 + edges (vertex names):
       6 1 -- 2, 3
       7 2 -- 1, 3, 3, 4
       8 3 -- 1, 2, 2, 5
       9 4 -- 2, 5, 6, 7
      10 5 -- 3, 4, 6
      11 6 -- 4, 5, 7
      12 7 -- 4, 6
      13 > is_simple(mg)
      14 [1] FALSE
```

Checking whether or not a network graph is simple is a somewhat trivial but nevertheless important preliminary step in doing a typical network analysis, as many models and methods assume the input graph to be simple or behave differently if it is not.

Note that it is straightforward, and indeed not uncommon in practice, to transform a multi-graph into a weighted graph, wherein each resulting proper edge is equipped with a weight equal to the multiplicity of that edge in the original multi-graph. For example, converting our toy multi-graph mg to a weighted graph results in a simple graph,

```
#2.26  1 > E(mg)$weight <- 1
       2 > wg2 <- simplify(mg)
       3 > is_simple(wg2)
       4 [1] TRUE
```

the edges of which match our initial toy graph g,

```
#2.27  1 > print_all(wg2)
       2 IGRAPH 78518b0 UNW- 7 10 -- Toy Graph
       3 + attr: name (g/c), name (v/c), color (v/c), weight
       4 | (e/n)
       5 + edges (vertex names):
       6 1 -- 2, 3
       7 2 -- 1, 3, 4
       8 3 -- 1, 2, 5
       9 4 -- 2, 5, 6, 7
      10 5 -- 3, 4, 6
      11 6 -- 4, 5, 7
      12 7 -- 4, 6
```

but for which the third edge (i.e., connecting vertices 2 and 3) has a weight of 2.

```
#2.28  1 > E(wg2)$weight
       2  [1] 1 1 2 1 1 1 1 1 1 1
```

Moving beyond such basic concerns regarding the nature of the edges in a graph, it is necessary to have a language for discussing the connectivity of a graph. The most basic notion of connectivity is that of adjacency. Two vertices $u, v \in V$ are said to be *adjacent* if joined by an edge in E. Such vertices are also referred to as *neighbors*. For example, the three neighbors of vertex 5 in our toy graph g are

```
#2.29  1  > neighbors(g,5)
       2  + 3/7 vertices, named, from 2e0a0da:
       3  [1] 3 4 6
```

Similarly, two edges $e_1, e_2 \in E$ are adjacent if joined by a common endpoint in V. A vertex $v \in V$ is *incident* on an edge $e \in E$ if v is an endpoint of e. From this follows the notion of the *degree* of a vertex v, say d_v, defined as the number of edges incident on v.

```
#2.30  1  > degree(g)
       2  1 2 3 4 5 6 7
       3  2 3 3 4 3 3 2
```

For digraphs, vertex degree is replaced by *in-degree* (i.e., d_v^{in}) and *out-degree* (i.e., d_v^{out}), which count the number of edges pointing in towards and out from a vertex, respectively.

```
#2.31  1  > degree(dg, mode="in")
       2   Sam Mary  Tom
       3    0    2    2
       4  > degree(dg, mode="out")
       5   Sam Mary  Tom
       6    2    1    1
```

It is also useful to be able to discuss the concept of movement about a graph. For example, a *walk* on a graph G, from v_0 to v_l, is an alternating sequence $\{v_0, e_1, v_1, e_2, \ldots, v_{l-1}, e_l, v_l\}$, where the endpoints of e_i are $\{v_{i-1}, v_i\}$. The *length* of this walk is said to be l. Refinements of a walk include *trails*, which are walks without repeated edges, and *paths*, which are trails without repeated vertices. A trail for which the beginning and ending vertices are the same is called a *circuit*. Similarly, a walk of length at least three, for which the beginning and ending vertices are the same, but for which all other vertices are distinct from each other, is called a *cycle*. Graphs containing no cycles are called *acyclic*. In a digraph, these notions generalize naturally. For example, a *directed walk* from v_0 to v_l proceeds from tail to head along arcs between v_0 and v_l.

A vertex v in a graph G is said to be *reachable* from another vertex u if there exists a walk from u to v. The graph G is said to be *connected* if every vertex is reachable from every other. A *component* of a graph is a maximally connected subgraph. That is, it is a connected subgraph of G for which the addition of any other remaining vertex in V would ruin the property of connectivity. The toy graph g, for example, is connected

```
#2.32  1  > is_connected(g)
       2  [1] TRUE
```

and therefore consists of only a single component

```
#2.33  1  > clusters(g)
       2  $`membership`
       3  1 2 3 4 5 6 7
       4  1 1 1 1 1 1 1
       5
       6  $csize
       7  [1] 7
       8
       9  $no
      10  [1] 1
```

For a digraph, there are two variations of the concept of connectedness. A digraph *G* is *weakly connected* if its underlying graph (i.e., the result of stripping away the labels 'tail' and 'head' from *G*) is connected. It is called *strongly connected* if every vertex *v* is reachable from every *u* by a directed walk. The toy graph dg, for example, is weakly connected but not strongly connected.

```
#2.34  1  > is_connected(dg,mode="weak")
       2  [1] TRUE
       3  > is_connected(dg,mode="strong")
       4  [1] FALSE
```

A common notion of *distance* between vertices on a graph is defined as the length of the shortest path(s) between the vertices (which we set equal to infinity if no such path exists). This distance is often referred to as *geodesic distance*, with 'geodesic' being another name for shortest paths. The value of the longest distance in a graph is called the *diameter* of the graph. Our toy graph g has diameter

```
#2.35  1  > diameter(g, weights=NA)
       2  [1] 3
```

Ultimately, the concepts above are only the most basic of graph-theoretic quantities. There are a wide variety of queries one might make about graphs and quantities to calculate as a part of doing descriptive network analysis. We cover more of these in Chap. 4.

2.4.2 Special Types of Graphs

Graphs come in all 'shapes and sizes,' as it were, but there are a number of families of graphs that are commonly encountered in practice. We illustrate this notion with the examples of four such families shown in Fig. 2.2.

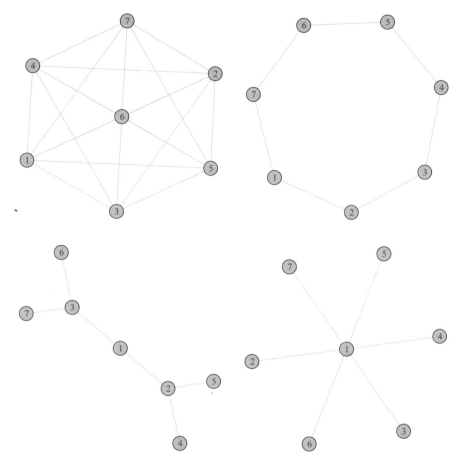

Fig. 2.2 Examples of graphs from four families. Complete (*top left*); ring (*top right*); tree (*bottom left*); and star (*bottom right*)

```
#2.36 1 > g.full <- make_full_graph(7)
      2 > g.ring <- make_ring(7)
      3 > g.tree <- make_tree(7, children=2, mode="undirected")
      4 > g.star <- make_star(7, mode="undirected")
      5 > par(mfrow=c(2, 2), mai = c(0.2, 0.2, 0.2, 0.2))
      6 > plot(g.full)
      7 > plot(g.ring)
      8 > plot(g.tree)
      9 > plot(g.star)
```

A *complete* graph is a graph where every vertex is joined to every other vertex by an edge. This concept is perhaps most useful in practice through its role in defining a *clique*, which is a complete subgraph. Shown in Fig. 2.2 is a complete graph of order $N_v = 7$, meaning that each vertex is connected to all of the other six vertices.

A *regular* graph is a graph in which every vertex has the same degree. A regular graph with common degree d is called d-*regular.* An example of a 2-regular graph is the ring shown in Fig. 2.2. The standard lattice, such as is associated visually with a checker board, is an example of a 4-regular graph.

A connected graph with no cycles is called a *tree*. The disjoint union of such graphs is called a *forest*. Trees are of fundamental importance in the analysis of networks. They serve, for example, as a key data structure in the efficient design of many computational algorithms. A digraph whose underlying graph is a tree is called a *directed tree*. Often such trees have associated with them a special vertex called a *root*, which is distinguished by being the only vertex from which there is a directed path to every other vertex in the graph. Such a graph is called a *rooted tree*. A vertex preceding another vertex on a path from the root is called an *ancestor*, while a vertex following another vertex is called a *descendant*. Immediate ancestors are called *parents*, and immediate descendants, *children*. A vertex without any children is called a *leaf*. The distance from the root to the farthest leaf is called the *depth* of the tree.

Given a rooted tree of this sort, it is not uncommon to represent it diagrammatically without any indication of its directedness, as this is to be understood from the definition of the root. Such a representation of a tree is shown in Fig. 2.2. Treating vertex 1 as the root, this is a tree of depth 2, wherein each vertex (excluding the leafs) is the ancestor of two descendants.

A *k-star* is a special case of a tree, consisting only of one root and k leaves. Such graphs are useful for conceptualizing a vertex and its immediate neighbors (ignoring any connectivity among the neighbors). A representation of a 7-star is given in Fig. 2.2.

An important generalization of the concept of a tree is that of a *directed acyclic graph* (i.e., the DAG). A DAG, as its name implies, is a graph that is directed and that has no directed cycles. However, unlike a directed tree, its underlying graph is not a tree, in that replacing the arcs with undirected edges leaves a (simple) graph that contains cycles. Our toy graph dg, for example, is directed but not a DAG

```
#2.37 1 > is_dag(dg)
       2 [1] FALSE
```

since the underlying graph is a triangle and hence a 3-cycle. Nevertheless, it is often possible to still design efficient computational algorithms on DAGs that take advantage of this near-tree-like structure.

Lastly, a *bipartite* graph is a graph $G = (V, E)$ such that the vertex set V may be partitioned into two disjoint sets, say V_1 and V_2, and each edge in E has one endpoint in V_1 and the other in V_2. Such graphs typically are used to represent 'membership' networks, for example, with 'members' denoted by vertices in V_1, and the corresponding 'organizations', by vertices in V_2. For example, they are popular in studying the relationship between actors and movies, where actors and movies play the roles of members and organizations, respectively.

Fig. 2.3 A bipartite network

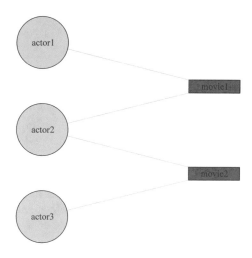

```
#2.38  1 > g.bip <- graph_from_literal(actor1:actor2:actor3,
       2 +    movie1:movie2, actor1:actor2 - movie1,
       3 +    actor2:actor3 - movie2)
       4 > V(g.bip)$type <- grepl("^movie", V(g.bip)$name)
       5 > print_all(g.bip, v=T)
       6 IGRAPH 68780ab UN-B 5 4 --
       7 + attr: name (v/c), type (v/l)
       8 + vertex attributes:
       9 |        name  type
      10 | [1] actor1 FALSE
      11 | [2] actor2 FALSE
      12 | [3] actor3 FALSE
      13 | [4] movie1  TRUE
      14 | [5] movie2  TRUE
      15 + edges from 68780ab (vertex names):
      16 [1] actor1--movie1 actor2--movie1 actor2--movie2
      17 [4] actor3--movie2
```

A visualization of g.bip is shown[9] in Fig. 2.3.

It is not uncommon to accompany a bipartite graph with at least one of two possible induced graphs. Specifically, a graph $G_1 = (V_1, E_1)$ may be defined on the vertex set V_1 by assigning an edge to any pair of vertices that both have edges in E to at least one common vertex in V_2. Similarly, a graph G_2 may be defined on V_2. Each of these graphs is called a *projection* onto its corresponding vertex subset. For example, the projection of the actor-movie network g.bip onto its two vertex subsets yields

---

[9]The R code for generating this visualization is provided in Chap. 3.

```
#2.39  1  > proj <- bipartite_projection(g.bip)
       2  > print_all(proj[[1]])
       3  IGRAPH 3782b1c UNW- 3 2 --
       4  + attr: name (v/c), weight (e/n)
       5  + edges from 3782b1c (vertex names):
       6  [1] actor1--actor2 actor2--actor3
       7  > print_all(proj[[2]])
       8  IGRAPH 3782b1c UNW- 2 1 --
       9  + attr: name (v/c), weight (e/n)
      10  + edge from 3782b1c (vertex names):
      11  [1] movie1--movie2
```

Within the actor network, `actor2` is adjacent to both `actor1` and `actor3`, as the former actor was in movies with each of the latter actors, although these latter were not themselves in any movies together, and hence do not share an edge. The movie network consists simply of a single edge defined by `movie1` and `movie2`, since these movies had actors in common.

2.5 Additional Reading

A more thorough introduction to the topic of graph theory may be found in any of a number of introductory textbooks, such as those by Bollobás [1], Diestel [3], or Gross and Yellen [4]. Details on graph data structures and algorithms are in many computer science algorithms texts. See the text by Cormen, Leiserson, Rivest, and Stein [2], for example.

References

1. B. Bollobás, *Modern Graph Theory*. New York: Springer, 1998.
2. T. Cormen, C. Leiserson, R. Rivest, and C. Stein, *Introduction to Algorithms*. Cambridge, MA: MIT Press, 2003.
3. R. Diestel, *Graph Theory, Third Edition*. Heidelberg: Springer-Verlag, 2005.
4. J. Gross and J. Yellen, *Graph Theory And Its Applications*. Boca Raton, FL: Chapman & Hall/CRC, 1999.
5. E. Lazega, *The Collegial Phenomenon: The Social Mechanisms of Cooperation Among Peers in a Corporate Law Partnership*. Oxford: Oxford University Press, 2001.

Chapter 3
Visualizing Network Data

3.1 Introduction

Up until this point, we have spoken only loosely of displaying network graphs, although we have shown several examples already. Here in this chapter we consider the problem of display in its own right. Techniques for displaying network graphs are the focus of the field of *graph drawing* or *graph visualization*. Such techniques typically seek to incorporate a combination of elements from mathematics, human aesthetics, and algorithms. After a brief characterization of the elements of graph visualization in Sect. 3.2, we look at a number of ways to lay out a graph, in Sect. 3.3, followed by some ways to further decorate such layouts, in Sect. 3.4. We also look quickly at some of the unique challenges posed by the problem of visualizing large network graphs in Sect. 3.5. Finally, in Sect. 3.6, we describe options for producing more sophisticated visualizations than those currently possible using R.

3.2 Elements of Graph Visualization

Suppose we have a set of network measurements that have been encoded in a network graph representation $G = (V, E)$, and we now wish to summarize G in a visual manner. At the heart of the graph visualization problem is the challenge of creating "geometric representations of . . . combinatorial structures," [5] using symbols (e.g., points, circles, squares, etc.) for vertices $v \in V$ and smooth curves for edges $e \in E$. For human consumption it is most convenient, of course, if a graph is drawn[1] in two-dimensional space, as opposed to three-dimensional space or on some more abstract surface. Hence, we will restrict our attention to this setting.

---

[1] Here and throughout we use terms like 'draw' only in the colloquial sense, although more formal mathematical treatments of this topic area exist (e.g., see Chap. 8 of Gross and Yellen [9]) which attach more specialized understandings to these terms.

© Springer Nature Switzerland AG 2020
E. D. Kolaczyk and G. Csárdi, *Statistical Analysis of Network Data with R*, Use R!,
https://doi.org/10.1007/978-3-030-44129-6_3

Intuitively, it is not hard to see that there are uncountably many ways that we could lay down a candidate set of points and curves on paper to represent a graph G. The important question for any such candidate, however, is whether or not it adequately communicates the desired relational information in G. While in principle this might suggest the drawing of graphs by hand, in practice hand drawings are only realistic for very small graphs. Generally graphs of nontrivial size must be drawn, at least in part, using automated methods.

In principle, one could simple lay down the vertices in G in a random fashion (e.g., uniformly) over a given region and then draw straight lines between those vertices connected by an edge. Unfortunately, although such drawings are simple enough to create, they tend to look like little more than a ball of yarn. In order to facilitate improved automatic drawings of graphs, various specifications or requirements have evolved—some firm, and some flexible—which have been formally categorized as drawing conventions, aesthetics, and constraints. See di Battista et al. [5] for a detailed treatment, or Kolaczyk [12, Sect. 3.4] for a brief summary.

From a practical perspective, drawing conventions, aesthetics, and constraints effectively serve to define parameters for automatic graph drawing methods, and the determination of a graph drawing frequently becomes a formal optimization over some or all of these parameters. Such optimizations typically are difficult to solve exactly in real time for graphs that are nontrivial in size. Therefore, it is common to develop computationally efficient algorithms that seek an approximate solution, often through the use of heuristics and the imposition of priorities among aesthetics.

Within **igraph**, the `plot` command calls a variety of such algorithms which, when applied to a graph object g, allows for the user to produce a fairly rich assortment of graph visualizations. The parameters associated with `plot` allow for both the specification of the algorithm to be used and the settings of various conventions, aesthetics, and constraints allowed for by a given algorithm.

3.3 Graph Layouts

At the heart of graph visualization is the graph layout, i.e., the placement of vertices and edges in space. There are far too many graph layout methods for us to present a full survey here. Rather, we discuss a handful of representative examples.

We will illustrate using two network graphs—a $5 \times 5 \times 5$ lattice

```
#3.1  1 > library(sand)
      2 > g.l <- make_lattice(c(5, 5, 5))
```

and a network of 'web-logs' or simply 'blogs'.

```
#3.2  1 > data(aidsblog)
      2 > summary(aidsblog)
      3 IGRAPH D--- 146 187 --
```

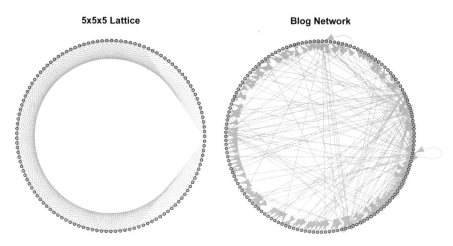

Fig. 3.1 Circular layouts

The blog network is a snapshot of the pattern of citation among 146 unique blogs related to AIDS, patients, and their support networks, collected by Gopal [7] over a randomly selected three-day period in August 2005. A directed edge from one blog to another indicates that the former has a link to the latter in their web page (more specifically, the former refers to the latter in their so-called 'blogroll').

Note that both graphs are of roughly the same order (125 and 146 vertices, respectively), although the former has almost twice as many edges as the latter (300 vs. 187). The lattice, however, is by definition highly uniform in its connectivity across vertices, whereas the blog network is not.

The simplest layout is a *circular layout*, wherein the vertices are arranged (usually equi-spaced) around the circumference of a circle. The edges are then drawn across the circle. Circular layouts of the lattice and blog networks are shown in Fig. 3.1.

```
#3.3 1 > igraph_options(vertex.size=3, vertex.label=NA,
     2 +    edge.arrow.size=0.5)
     3 > par(mfrow=c(1, 2))
     4 > plot(g.l, layout=layout_in_circle)
     5 > title("5x5x5 Lattice")
     6 > plot(aidsblog, layout=layout_in_circle)
     7 > title("Blog Network")
```

The visualization of the lattice is much more pleasing to the eye than that of the blog network, largely due to the low level of edge-crossings through the center of the circle. Ordering of the vertices around the circle is important with this type of layout—a random re-ordering of the vertices in the lattice, for example, would yield a picture much more like that of the blog network. Common vertex orderings for circular layouts include ordering by degree and grouping by common vertex attributes.

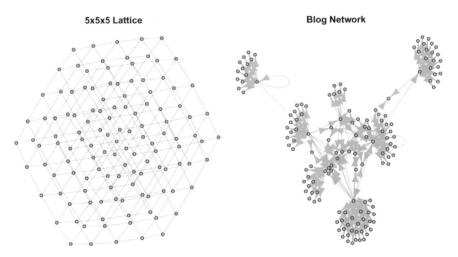

Fig. 3.2 Layouts using the method of Fruchterman and Reingold

Often more effective for creating useful drawings are layouts based on exploiting analogies between the relational structure in graphs and the forces among elements in physical systems. One approach in this area, and the earliest proposed, is to introduce attractive and repulsive forces by associating vertices with balls and edges with springs. If a literal system of balls connected by springs is disrupted, thereby stretching some of the springs and compressing others, upon being let go it will return to its natural state. So-called *spring-embedder* methods of graph drawing define a notion of force for each vertex in the graph depending, at the very least, on the positions of pairs of vertices and the distances between them, and seek to iteratively update the placement of vertices until a vector of net forces across vertices converges.

The method of Fruchterman and Reingold [6] is a commonly used example of this type. Applied to the lattice and blog networks,

```
#3.4  1 > plot(g.l,layout=layout_with_fr)
      2 > title("5x5 Lattice")
      3 > plot(aidsblog,layout=layout_with_fr)
      4 > title("Blog Network")
```

as shown in Fig. 3.2, we see that substantially more of the structure inherent to each network is now visible.

Alternatively, motivated by the fact that it is possible to associate the collection of forces in spring systems with an overall system energy, another common approach to generating layouts is that of *energy-placement* methods. An energy, as a function of vertex positions, ostensibly is defined using expressions motivated by those found in physics. A vertex placement is chosen which minimizes the total system energy. A physical system with minimum energy is typically in its most relaxed state, and hence the assertion here is that a graph drawn according to similar principles should be visually appealing.

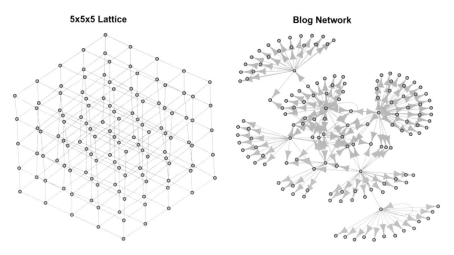

Fig. 3.3 Layouts using the method of Kamada and Kawai

Methods based on multidimensional scaling (MDS), which have a long history in the social network literature, are of this type. The method of Kamada and Kawai [10] is a popular variant. Using this layout,

```
#3.5  1 > plot(g.l, layout=layout_with_kk)
      2 > title("5x5x5 Lattice")
      3 > plot(aidsblog, layout=layout_with_kk)
      4 > title("Blog Network")
```

the resulting visualizations of our lattice and blog networks are similar in spirit to those obtained using Fruchterman–Reingold. See Fig. 3.3.

In some cases, network graphs have special structure that it is desirable to accentuate. Trees are one such case. Consider, for example, the visualizations in Fig. 3.4.

```
#3.6  1 > g.tree <- graph_from_literal(1-+2,1-+3,1-+4,2-+5,2-+6,
      2 +                               2-+7,3-+8,3-+9,4-+10)
      3 > par(mfrow=c(1, 3))
      4 > igraph_options(vertex.size=30, edge.arrow.size=0.5,
      5 +     vertex.label=NULL)
      6 > plot(g.tree, layout=layout_in_circle)
      7 > plot(g.tree, layout=layout_as_tree(g.tree, circular=T))
      8 > plot(g.tree, layout=layout_as_tree)
```

With the circular layout, it is not obvious that the graph is a tree. However, with both the radial layout—in which edges radiate outward on concentric circles—and the layered layout, the structure of the graph is immediately apparent.

Similarly, bipartite graphs often are laid out with the two sets of vertices running across opposing rows (or down opposing columns)—one type of vertex within each row (column)—and with edges running between the rows (columns), in a manner reminiscent of railroad tracks. The visualization shown in Fig. 2.3, for example, is of this sort.

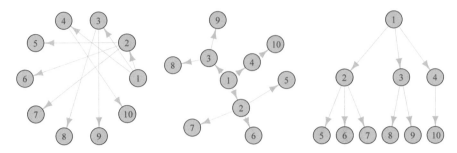

Fig. 3.4 Three layouts of the same tree: circular (*left*), radial (*center*), and layered (*right*)

```
#3.7 1 > plot(g.bip, layout= -layout_as_bipartite(g.bip)[,2:1],
     2 +    vertex.size=60, vertex.shape=ifelse(V(g.bip)$type,
     3 +        "rectangle", "circle"),
     4 +    vertex.label.cex=1.75,
     5 +    vertex.color=ifelse(V(g.bip)$type, "red", "cyan"))
```

3.4   Decorating Graph Layouts

While the relative positions of vertices and the placement of edges between them is clearly important in visualizing network graphs, additional network information—when available—can be incorporated into visualizations by varying characteristics like the size, shape, and color of vertices and edges. In particular, such techniques allow for the visualization of decorated graphs.

Consider, for example, the so-called 'karate club network' of Zachary [16]. Nodes represent members of a karate club observed by Zachary for roughly 2 years during the 1970s, and links connecting two nodes indicate social interactions between the two corresponding members. This dataset is somewhat unique in that Zachary had the curious fortune (from a scientific perspective) to witness the club split into two different clubs during his period of observation, due to a dispute between the head teacher and an administrator. Attribute information available for this network includes identification of the head teacher and the administrator, membership in one of the two factions underlying the eventual split, and relative frequency of interactions between members. While clearly more involved than just issuing a simple call to `plot()`, it is nevertheless straightforward to incorporate all of this information into a visualization of this network.

```
#3.8  1 > library(igraphdata)
      2 > data(karate)
      3 > # Reproducible layout
      4 > set.seed(42)
      5 > l <- layout_with_kk(karate)
      6 > # Plot undecorated first.
      7 > igraph_options(vertex.size=10)
      8 > par(mfrow=c(1,1))
      9 > plot(karate, layout=l, vertex.label=V(karate),
     10 +    vertex.color=NA)
```

```
11 > # Now decorate, starting with labels.
12 > V(karate)$label <- sub("Actor ", "", V(karate)$name)
13 > # Two leaders get shapes different from club members.
14 > V(karate)$shape <- "circle"
15 > V(karate)[c("Mr Hi", "John A")]$shape <- "rectangle"
16 > # Differentiate two factions by color.
17 > V(karate)[Faction == 1]$color <- "red"
18 > V(karate)[Faction == 2]$color <- "dodgerblue"
19 > # Vertex area proportional to vertex strength
20 > # (i.e., total weight of incident edges).
21 > V(karate)$size <- 4*sqrt(strength(karate))
22 > V(karate)$size2 <- V(karate)$size * .5
23 > # Weight edges by number of common activities
24 > E(karate)$width <- E(karate)$weight
25 > # Color edges by within/between faction.
26 > F1 <- V(karate)[Faction==1]
27 > F2 <- V(karate)[Faction==2]
28 > E(karate)[ F1 %--% F1 ]$color <- "pink"
29 > E(karate)[ F2 %--% F2 ]$color <- "lightblue"
30 > E(karate)[ F1 %--% F2 ]$color <- "yellow"
31 > # Offset vertex labels for smaller points (default=0).
32 > V(karate)$label.dist <-
33 +   ifelse(V(karate)$size >= 9.0, 0, 1.0)
34 > # Plot decorated graph, using same layout.
35 > plot(karate, layout=l)
```

The resulting visualization is shown in Fig. 3.5. Also shown, for comparison, is a visualization using the same layout coordinates (generated according to the Kamada–Kawai algorithm), but without any decoration. A substantial amount of additional information is communicated by way of the decorated visualization, where vertices are sized in proportion to their (weighted) degree, the relative frequency of interactions is shown using edge thickness, a change in vertex shape indicates the faction leaders, and colors are used to distinguish membership in the factions as well as edges joining within versus between the same factions.

A similar use of vertex color, shape, and size was used in the visualization of the Lazega lawyer network in Fig. 1.1. Note that within **igraph**, supplying the graphical parameters to the plotting command is an alternative to setting them as vertex and edge attributes. We demonstrate below.

```
#3.9  1 > library(sand)
      2 > data(lazega)
      3 > # Office location indicated by color.
      4 > colbar <- c("red", "dodgerblue", "goldenrod")
      5 > v.colors <- colbar[V(lazega)$Office]
      6 > # Type of practice indicated by vertex shape.
      7 > v.shapes <- c("circle", "square")[V(lazega)$Practice]
      8 > # Vertex size proportional to years with firm.
      9 > v.size <- 3.5*sqrt(V(lazega)$Years)
     10 > # Label vertices according to seniority.
     11 > v.label <- V(lazega)$Seniority
     12 > # Reproducible layout.
     13 > set.seed(42)
     14 > l <- layout_with_fr(lazega)
```

```
15 > plot(lazega, layout=1, vertex.color=v.colors,
16 +    vertex.shape=v.shapes, vertex.size=v.size,
17 +    vertex.label=v.label)
```

3.5   Visualizing Large Networks

Despite their sophistication, for all of the methods described so far, the graph draw-ings will tend to look increasingly cluttered as the number of vertices N_v nears 100 or so—and simply unintelligible for thousands of vertices or more—due to the finiteness of the available space and resolution.

For example, in Fig. 3.6 is shown a visualization of a subnetwork of French polit-ical blogs, extracted from a snapshot of over 1,100 such blogs on a single day in October of 2006 and classified by the "Observatoire Presidentielle" project as to political affiliation.[2] The network consists of 192 blogs linked by 1,431 edges, the latter indicating that at least one of the two blogs referenced the other.

```
#3.10 1 > library(sand)
      2 > summary(fblog)
      3 IGRAPH ee79c98 UN-- 192 1431 --
      4 + attr: name (v/c), PolParty (v/c)
```

Nine political parties are represented among these blogs, and have been included as vertex attributes.

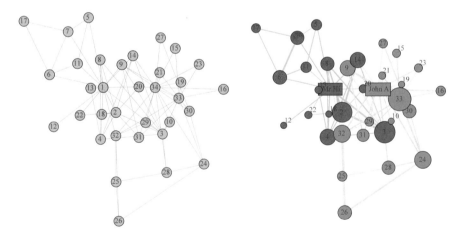

Fig. 3.5 Plain (*left*) and decorated (*right*) visualizations of Zachary's karate club network

---

[2]Original source: http://observatoire-presidentielle.fr/. The subnetwork used here is part of the **mixer** package in R. Note that the inherent directionality of blogs are ignored in these data, as the network graph is undirected.

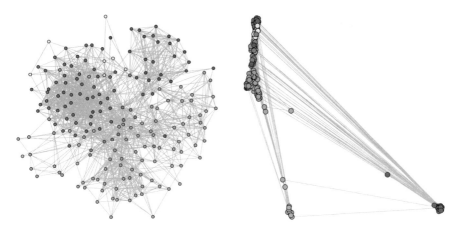

Fig. 3.6 Visualizations of the French political blog network using layouts generated by the Kamada–Kawai (*left*) and DrL (*right*) algorithms

```
#3.11  1 > party.names <- sort(unique(V(fblog)$PolParty))
       2 > party.names
       3 [1] " Cap21"                " Commentateurs Analystes"
       4 [3] " Les Verts"            " liberaux"
       5 [5] " Parti Radical de Gauche" " PCF - LCR"
       6 [7] " PS"                   " UDF"
       7 [9] " UMP"
```

The visualization shown on the left in Fig. 3.6 was produced using a layout generated by the standard Kamada–Kawai method.

```
#3.12  1 > set.seed(42)
       2 > l = layout_with_kk(fblog)
       3 > party.nums.f <- as.factor(V(fblog)$PolParty)
       4 > party.nums <- as.numeric(party.nums.f)
       5 > # igraph color palette has 8 colors
       6 > # 9 colors needed (for 9 political parties)
       7 > library(RColorBrewer)
       8 > colrs <- brewer.pal(9,"Set1")
       9 > V(fblog)$color <- colrs[party.nums]
      10 > plot(fblog, layout=l, vertex.label=NA,
      11 +        vertex.size=3)
```

Note that, while it is indeed possible, with a bit of effort and the aid of color indicating political party affiliation, to distinguish some structure, the plot is nevertheless rather 'busy'.

Fortunately, there are layout algorithms designed specifically for the purpose of visualizing large networks. For example, VxOrd [4], a visualization package produced by Sandia Labs, is an enhanced version of the spring-embedder methodology. It attempts to place vertices in clusters on the two-dimensional plane, with the help of sophisticated optimization methods to more efficiently search the space of possible graph drawings and a grid that helps reduce computation time from the typical $O(N_v^2)$ down to $O(N_v)$. In addition, it employs edge-cutting criteria, designed towards producing drawings that balance the detail with which both local and global structure are shown. VxOrd has been used by Boyack et al. [1] to produce a map of the 'backbone' of Science, involving over 7,000 vertices (journals) and 16 million edges (co-citations). See also [12, Sect. 3.5.1] for a short summary of their analysis.

The DrL method [13], which is based on VxOrd, has been implemented in **igraph**. Applied to the French political blog data

```
#3.13  1 > set.seed(42)
       2 > l <- layout_with_drl(fblog)
       3 > plot(fblog, layout=l, vertex.size=5, vertex.label=NA)
```

and shown on the right in Fig. 3.6, we find that the method has clustered blogs in a way that is strongly influenced by certain of the party affiliations, despite not having had this information.[3]

When such clustering exists and may be characterized explicitly, either because it occurs with respect to a measured variable or because it is inferred through so-called graph partitioning methods,[4] it may be useful to coarsen a network graph prior to visualization, replacing groups of vertices with single meta-vertices. This is demonstrated on the French political blog data in Fig. 3.7. Having first coarsened the network, by aggregating edges between groups,

```
#3.14  1 > fblog.c <- contract(fblog, party.nums)
       2 > E(fblog.c)$weight <- 1
       3 > fblog.c <- simplify(fblog.c)
```

we plot the resulting network.

```
#3.15  1 > party.size <- as.vector(table(V(fblog)$PolParty))
       2 > plot(fblog.c, vertex.size=5*sqrt(party.size),
       3 +        vertex.label=party.names, vertex.color=colrs,
       4 +        edge.width=sqrt(E(fblog.c)$weight),
       5 +        vertex.label.dist=3.5, edge.arrow.size=0)
```

In the resulting visualization, the size of the groups defined by political parties in the original network, and the numbers of edges between those groups, are reflected in vertex size and edge thickness, respectively. The relationships among political parties that was only suggested by the visualizations in Fig. 3.6 is now quite evident.

Alternatively, specific information we desire to be communicated through a network visualization might suggest that only a relevant subgraph(s) be shown. For example, sometimes it is useful to highlight the structure local to a given vertex, such as in the so-called *egocentric* network visualizations commonly used in the social network literature, which show the vertex, its immediate neighbors, and all edges among them.

---

[3]While the DrL algorithm has been found to scale successfully to networks of over 1 million vertices, it is known to produce less than satisfactory results for networks with only hundreds of vertices. Our use of a smaller network here is primarily for the purpose of illustration.

[4]See Sect. 4.4.

Fig. 3.7 Visualization of the French political blog network at the level of political parties

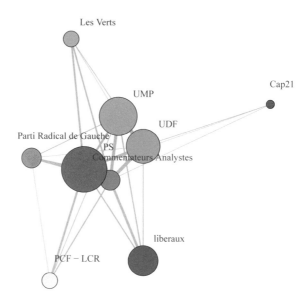

As an illustration, consider again the karate network. Extracting the (first-order) neighborhoods surrounding each vertex

```
#3.16 1 > data(karate)
      2 > k.nbhds <- make_ego_graph(karate, order=1)
```

we see, for example, that the neighborhoods pertaining to the head instructor (Mr Hi, vertex 1) and administrator (John A, vertex 34) are the largest.

```
#3.17 1 > sapply(k.nbhds, vcount)
      2   [1] 17 10 11  7  4  5  5  5  6  3  4  2  3  6  3  3  3
      3  [18]  3  3  4  3  3  3  6  4  4  3  5  4  5  5  7 13 18
```

Pulling out these two largest subnetworks and plotting them,

```
#3.18 1 > k.1 <- k.nbhds[[1]]
      2 > k.34 <- k.nbhds[[34]]
      3 > par(mfrow=c(1,2))
      4 > plot(k.1, vertex.label=NA,
      5 +     vertex.color=c("red", rep("lightblue", 16)))
      6 > plot(k.34, vertex.label=NA,
      7 +     vertex.color=c(rep("lightblue", 17), "red"))
```

we obtain the visualizations shown in Fig. 3.8. Comparing these plots to that of the full karate network, as seen in Fig. 3.5, it is clear that these two subnetworks capture the vast majority of the structure in the full network.

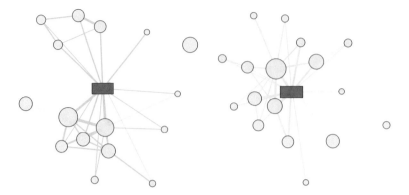

Fig. 3.8 Two ego-centric views of the karate club network of Fig. 3.5, from the perspectives of vertices 1 (*left*) and 34 (*right*), the authority figures (instructor and administrator, respectively, shown in *red*) about which the club ultimately split. (Vertex labels have been dropped, for the purposes of this illustration, to reduce clutter.)

3.6 Using Visualization Tools Outside of R

While using R for visualizing networks has the advantage of a programmatical inter-face, high quality graphics, and many graphics file formats, the network drawing itself is somewhat limited. For example, there are relatively few choices available for node and arrow shapes, and label placement is somewhat awkward, so that avoiding overlapping nodes and labels is not easy. Interactive editing of graph drawings is even more limited. However, if producing a network visualization of particularly high quality is a goal, there are certainly other software tools for this job. To close this chapter, we briefly touch upon some of these options, and discuss how data can be passed to them from R.

Graphviz is a classic graph layout and drawing tool. It is a collection of stan-dalone command line programs, with a separately developed interface to R, via the **Rgraphviz** package. An easy way to use an **igraph** graph with Graphviz is to write it to a file in the 'dot' file format that is supported by **igraph**'s `write_graph` function.

Pajek is one of the first network analysis and visualization tools. It is a standalone program for Microsoft Windows environments, and supports interactive visualiza-tion and editing as well. Pajek has its own file format, that is supported by **igraph**'s `write_graph` and `read_graph` functions (with some limitations), so transfer-ring data between R and Pajek is usually easy.

Cytoscape is a modern cross-platform network analysis and visualization tool, written in Java, specializing in biological networks. It has several layout algorithms implemented and sophisticated interactive visualization, and it is under active devel-opment. Cytoscape can read and write the GML file format, also supported by **igraph**. In addition, it supports simple tables in text files, which is advantageous if the graph has many attributes.

Gephi is another standalone cross-platform tool for analysis and visualization of graphs, with sophisticated interactive editing. It has its own file format called 'gexf', and the **rgexf** package is able to convert 'gexf' to and from **igraph** graphs.

There are several other tools available, of course. These usually support some of the file formats R can read and write (e.g. GraphML, GML, CSV, etc.), so network data can be passed around easily. Visualization of vertex and edge attributes are sometimes more problematic, however.

3.7 Additional Reading

An extensive overview of principles and techniques for network graph visualization may be found in the edited volume of Kaufmann and Wagner [11]. For detailed development of many of the corresponding algorithms, see the text of di Battista et al. [5]. Finally, for treatment of the more established body of work on the visualization of data in general, see the classic books by Tukey [15], Tufte [14], and Cleveland [2, 3], for example, or any number of more recent volumes, such as Grant [8].

References

1. K. Boyack, R. Klavans, and K. Börner, "Mapping the backbone of Science," *Scientometrics*, vol. 64, no. 3, pp. 351–374, 2005.
2. W. Cleveland, *The Elements of Graphing Data*. Wadsworth, 1985.
3. W. Cleveland, *Visualizing Data*. Hobart Press, 1993.
4. G. Davidson, B. Wylie, and K. Boyack, "Cluster stability and the use of noise in interpretation of clustering," *IEEE Symposium on Information Visualization*, pp. 23–30, 2002. 5
5. G. di Battista, P. Eades, R. Tamassia, and I. Tollis, *Graph Drawing*. Englewood Cliffs, NJ: Prentice Hall, 1999.
6. T. Fruchterman and E. Reingold, "Graph drawing by force-directed placement," *Software–Practice and Experience*, vol. 21, no. 11, pp. 1129–1164, 1991.
7. S. Gopal, "The evolving social geography of blogs," in *Societies and Cities in the Age of Instant Access*, H. Miller, Ed. Berlin: Springer, 2007, pp. 275–294.
8. R. Grant, *Data Visualization: Charts, Maps, and Interactive Graphics*. Chapman and Hall/CRC, 2018.
9. J. Gross and J. Yellen, *Graph Theory And Its Applications*. Boca Raton, FL: Chapman & Hall/CRC, 1999.
10. T. Kamada and S. Kawai, "An algorithm for drawing general undirected graphs," *Information Processing Letters*, vol. 31, no. 1, pp. 7–15, 1989.
11. M. Kaufmann and D. Wagner, Eds., *Drawing Graphs*. Berlin: Springer, 1998.
12. E. Kolaczyk, *Statistical Analysis of Network Data: Methods and Models*. Springer Verlag, 2009.
13. S. Martin, W. Brown, R. Klavans, and K. Boyack, "DrL: Distributed recursive (graph) layout," SAND Reports, Tech. Rep. 2936, 2008.
14. E. Tufte, *The Visual Display of Quantitative Information*. Graphics Press, 1983.
15. J. Tukey, *Exploratory Data Analysis*. New York: Addison-Wesley, 1977.
16. W. Zachary, "An information flow model for conflict and fission in small groups," *Journal of Anthropological Research*, vol. 33, no. 4, pp. 452–473, 1977.

Chapter 4
Descriptive Analysis of Network Graph Characteristics

4.1 Introduction

In the study of a given complex system, questions of interest can often be re-phrased in a useful manner as questions regarding some aspect of the structure or characteristics of a corresponding network graph. For example, various types of basic social dynamics can be represented by triplets of vertices with a particular pattern of ties among them (i.e., triads); questions involving the movement of information or commodities usually can be posed in terms of paths on the network graph and flows along those paths; certain notions of the 'importance' of individual system elements may be captured by measures of how 'central' the corresponding vertex is in the network; and the search for 'communities' and analogous types of unspecified 'groups' within a system frequently may be addressed as a graph partitioning problem.

The structural analysis of network graphs has traditionally been treated primarily as a descriptive task, as opposed to an inferential task, and the tools commonly used for such purposes derive largely from areas outside of 'mainstream' statistics. For example, an overwhelming proportion of these tools are naturally graph-theoretic in nature, and thus have their origins in mathematics and computer science. Similarly, the field of social network analysis has been another key source, contributing tools usually aimed—at least originally—at capturing basic aspects of social structure and dynamics. More recently, the field of physics has also been an important contributor, with the proposed tools often motivated by analogues in statistical mechanics.

We present in this chapter a brief overview of some of the many such tools available, starting with summaries of vertex and edge characteristics, in Sect. 4.2, continuing with measures of network cohesion, in Sect. 4.3, and with methods for graph partitioning (aka 'community detection'), in Sect. 4.4, and finishing with the topics of assortativity and mixing, in Sect. 4.5.

© Springer Nature Switzerland AG 2020
E. D. Kolaczyk and G. Csárdi, *Statistical Analysis of Network Data with R*, Use R!,
https://doi.org/10.1007/978-3-030-44129-6_4

4.2   Vertex and Edge Characteristics

As the fundamental elements of network graphs are their vertices and edges, there are a number of network characterizations centered upon these. We discuss several such characterizations here in this section. Our presentation is broken down according to (i) those characterizations based upon vertex degrees, and (ii) those seeking to capture some more general notion of the 'importance' of a vertex—typically referred to as vertex centrality measures. We also discuss the extension of such concepts from vertices to edges.

4.2.1   Vertex Degree

Recall that the degree d_v of a vertex v, in a network graph $G = (V, E)$, counts the number of edges in E incident upon v. Given a network graph G, define f_d to be the fraction of vertices $v \in V$ with degree $d_v = d$. The collection $\{f_d\}_{d \geq 0}$ is called the *degree distribution* of G, and is simply a rescaling of the set of degree frequencies, formed from the original degree sequence.

The degree distribution for the karate club network is shown in Fig. 4.1, using a histogram.

```
#4.1  1 > library(sand)
      2 > data(karate)
      3 > hist(degree(karate), col="lightblue", xlim=c(0,50),
      4 +    xlab="Vertex Degree", ylab="Frequency", main="")
```

We can see that there are three distinct groups of vertices, as measured by degree. The two most highly connected vertices correspond to actors 1 and 34 in the network, representing the instructor and administrator about whom the club eventually split. The next set of vertices consists of actors 2, 3, and also 33. Examination of our visualization of this network in Fig. 3.5 shows these actors to be the closest to actors 1 and 34, respectively.

For weighted networks, a useful generalization of degree is the notion of vertex *strength*, which is obtained simply by summing up the weights of edges incident to a given vertex. The distribution of strength—sometimes called the weighted degree distribution—is defined in analogy to the ordinary degree distribution. To illustrate, vertex strength for the karate club network is also shown in Fig. 4.1.

```
#4.2  1 > hist(strength(karate), col="pink",
      2 +    xlab="Vertex Strength", ylab="Frequency", main="")
```

For this network, the range of vertex strength stretches well beyond that of vertex degree, and the previously observed distinction among the three groups of vertices is lost. Nevertheless, both histograms arguably communicate important information about the network.

Degree distributions can exhibit a variety of shapes. For a network of interactions among protein pairs in yeast, for example,

Fig. 4.1 Distributions of vertex degree (*top*) and strength (*bottom*) for the karate club network

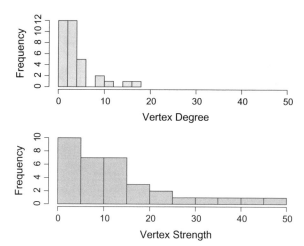

```
#4.3 1 > library(igraphdata)
     2 > data(yeast)
```

the shape is quite different from that of the karate club. In this case, the distribution of degrees associated with the

```
#4.4 1 > ecount(yeast)
     2 [1] 11855
```

edges among

```
#4.5 1 > vcount(yeast)
     2 [1] 2617
```

vertices is quite heterogeneous, as can be seen from examination of the histogram in Fig. 4.2.

```
#4.6 1 > d.yeast <- degree(yeast)
     2 > hist(d.yeast,col="blue",
     3 +    xlab="Degree", ylab="Frequency",
     4 +    main="Degree Distribution")
```

In particular, while there is a substantial fraction of vertices of quite low degree, of an order of magnitude similar to those of the karate network, there are also a nontrivial number of vertices with degrees at successively higher orders of magnitude.

Given the nature of the decay in this distribution, a log–log scale is more effective in summarizing the degree information.

```
#4.7 1 > dd.yeast <- degree_distribution(yeast)
     2 > d <- 1:max(d.yeast)-1
     3 > ind <- (dd.yeast != 0)
     4 > plot(d[ind], dd.yeast[ind], log="xy", col="blue",
     5 +    xlab=c("Log-Degree"), ylab=c("Log-Intensity"),
     6 +    main="Log-Log Degree Distribution")
```

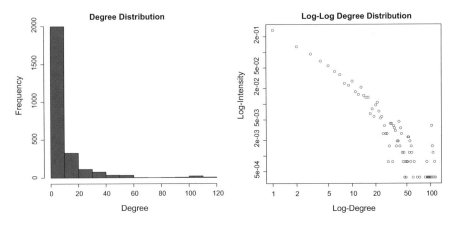

Fig. 4.2 Degree distribution for a network of protein–protein interactions in yeast. *Left*: Original scale; *Right*: log–log scale

From the plot on the right in Fig. 4.2 we see that there is a fairly linear decay in the log-frequency as a function of log-degree. While it is tempting to summarize the rate of this decay (i.e., the so-called degree exponent) using, for example, a simple linear regression, more sophisticated methods are preferable here. See [16, Sect. 4.2.1.1].

Beyond the degree distribution itself, it can be interesting to understand the manner in which vertices of different degrees are linked with each other. Useful in assessing this characteristic is the notion of the average degree of the *neighbors* of a given vertex. For example, a plot of average neighbor degree versus vertex degree in the yeast data, as shown in Fig. 4.3, suggests that while there is a tendency for vertices of higher degrees to link with similar vertices, vertices of lower degree tend to link with vertices of both lower and higher degrees.

```
#4.8 1 > a.nn.deg.yeast <- knn(yeast,V(yeast))$knn
     2 > plot(d.yeast, a.nn.deg.yeast, log="xy",
     3 +    col="goldenrod", xlab=c("Log Vertex Degree"),
     4 +    ylab=c("Log Average Neighbor Degree"))
```

4.2.2 Vertex Centrality

Many questions that might be asked about a vertex in a network graph essentially seek to understand its 'importance' in the network. Which actors in a social network seem to hold the 'reins of power'? How authoritative does a particular page in the World Wide Web seem to be considered? The deletion of which genes in a gene regulatory network is likely to be lethal to the corresponding organism? How critical is a given router in an Internet network to the flow of traffic? Measures of *centrality* are designed

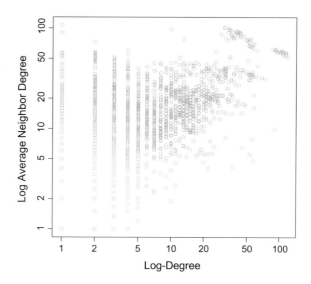

Fig. 4.3 Average neighbor degree versus vertex degree (log–log scale) for the yeast data

to quantify such notions of 'importance' and thereby facilitate the answering of such questions.

There are a vast number of different centrality measures that have been proposed over the years. We have already encountered what is arguably the most widely used measure of vertex centrality: vertex degree. Here we will focus our discussion primarily around the most common versions of three other classic types of vertex centrality measures—typically termed closeness, betweenness, and eigenvector centrality, respectively.

Closeness centrality measures attempt to capture the notion that a vertex is 'central' if it is 'close' to many other vertices. The standard approach, introduced by Sabidussi [21], is to let the centrality vary inversely with a measure of the total distance of a vertex from all others,

$$c_{Cl}(v) = \frac{1}{\sum_{u \in V} \mathsf{dist}(v, u)} \ , \tag{4.1}$$

where $\mathsf{dist}(v, u)$ is the geodesic distance between the vertices $u, v \in V$. Often, for comparison across graphs and with other centrality measures, this measure is normalized to lie in the interval $[0, 1]$, through multiplication by a factor $N_v - 1$.

Betweenness centrality measures are aimed at summarizing the extent to which a vertex is located 'between' other pairs of vertices. These centralities are based upon the perspective that 'importance' relates to where a vertex is located with respect to the paths in the network graph. If we picture those paths as the routes by which, say, communication of some sort or another takes place, vertices that sit on many paths are likely more critical to the communication process. The most commonly used betweenness centrality, introduced by Freeman [10], is defined as

$$c_B(v) = \sum_{s \neq t \neq v \in V} \frac{\sigma(s, t|v)}{\sigma(s, t)} \quad , \tag{4.2}$$

where $\sigma(s, t|v)$ is the total number of shortest paths between s and t that pass through v, and $\sigma(s, t)$ is the total number of shortest paths between s and t (regardless of whether or not they pass through v). In the event that shortest paths are unique, $c_B(v)$ just counts the number of shortest paths going through v. This centrality measure can be restricted to the unit interval through division by a factor of $(N_v - 1)(N_v - 2)/2$.

Finally, other centrality measures are based on notions of 'status' or 'prestige' or 'rank.' That is, they seek to capture the idea that the more central the neighbors of a vertex are, the more central that vertex itself is. These measures are inherently implicit in their definition and typically can be expressed in terms of eigenvector solutions of appropriately defined linear systems of equations. There are many such *eigenvector centrality* measures. For example, Bonacich [1], following work of Katz [14] and others, defined a centrality measure of the form

$$c_{Ei}(v) = \alpha \sum_{\{u,v\} \in E} c_{Ei}(u) \ . \tag{4.3}$$

The vector $\mathbf{c}_{Ei} = (c_{Ei}(1), \dots, c_{Ei}(N_v))^T$ is the solution to the eigenvalue problem $\mathbf{A}\mathbf{c}_{Ei} = \alpha^{-1}\mathbf{c}_{Ei}$, where $\mathbf{A}$ is the adjacency matrix for the network graph G. Bonacich [1] argues that an optimal choice of α^{-1} is the largest eigenvalue of $\mathbf{A}$, and hence $\mathbf{c}_{Ei}$ is the corresponding eigenvector. When G is undirected and connected, the largest eigenvalue of $\mathbf{A}$ will be simple and its eigenvector will have entries that are all nonzero and share the same sign. Convention is to report the absolute values of these entries, which will automatically lie between 0 and 1 by the orthonormality of eigenvectors.

An intuitively appealing way of displaying vertex centralities (for networks of small to moderate size) is to use a radial layout, with more central vertices located closer to the center. The function gplot.target, in the package **sna**, can be used for this purpose. For example,

```
#4.9 1 > A <- as_adjacency_matrix(karate, sparse=FALSE)
     2 > library(network)
     3 > g <- network::as.network.matrix(A)
     4 > library(sna)
     5 > sna::gplot.target(g, degree(g,gmode="graph"),
     6 +     main="Degree", circ.lab = FALSE,
     7 +     circ.col="skyblue", usearrows = FALSE,
     8 +     vertex.col=c("blue", rep("red", 32), "yellow"),
     9 +     edge.col="darkgray")
```

produces a visualization of degree centrality for the karate club network, as shown in Fig. 4.4. The visualizations of the other three centralities shown in this figure are produced similarly, replacing the argument degree(g) by the arguments closeness(g), betweenness(g), and evcent(g), respectively.[1]

---

[1]Note that **igraph** and **sna** use different formats for storing network data objects. Here we handle the conversion by first producing an adjacency matrix for the karate club object karate. The two

The plots in Fig. 4.4 illustrate how these four measures of centrality indeed capture different notions of 'central'. Note that the relative centralities of the two arguably most important figures in this club—the head teacher and the administrator—vary with choice of measure, from nearly indistinct under degree centrality to most distinct under betweenness centrality. Furthermore, under the latter choice of centrality, the

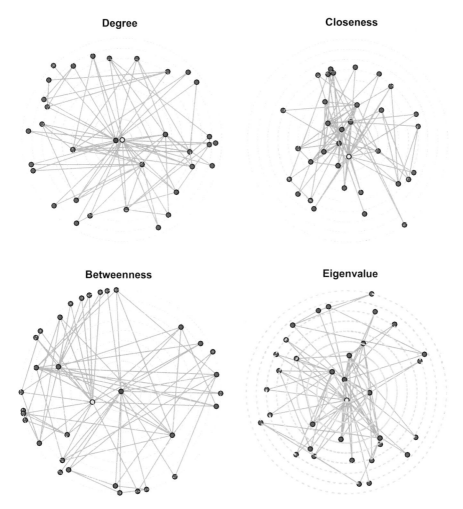

Fig. 4.4 Target plots showing various vertex centralities for the karate club network. The head teacher and instructor are indicated with *blue* and *yellow*, respectively

packages do, however, share several similar naming conventions. For example, to produce the analogous output of the **sna** functions `degree`, `closeness`, `betweenness`, and `evcent`, respectively, with the argument `g`, use the **igraph** functions `degree`, `closeness`, `betweenness`, and `eigen_centrality` with the argument `karate` (and, of course, appropriate choice of auxiliary arguments).

other actors in the network appear to be maximally distinct from those two as well. This last characteristic is a function of the largely star-like nature of this network, in which relatively few vertices v have shortest paths between pairs of other vertices s and t passing through v.

Extensions of these centrality measures from undirected to directed graphs are straightforward. However, in the latter case, there are in addition other useful options. For example, the HITS algorithm, based on the concept of 'hubs and authorities,' as introduced by Kleinberg [15] in the context of the World Wide Web, characterizes the importance of so-called hub vertices by how many authority vertices they point to, and so-called authority vertices by how many hubs point to them. Specifically, given an adjacency matrix $\mathbf{A}$ for a directed graph, hubs are determined according to the eigenvector centrality of the matrix $\mathbf{M}_{hub} = \mathbf{A}\mathbf{A}^{T}$, and authorities, according to that of $\mathbf{M}_{auth} = \mathbf{A}^{T}\mathbf{A}$.

Applying these measures to the AIDS blog network indicates, as seen in Fig. 4.5, that only six of the 146 blogs in this network play the role of a hub, while the vast majority of the vertices (including some of the hubs) play the role of an authority.

```
#4.10  1  > l <- layout_with_kk(aidsblog)
       2  > plot(aidsblog, layout=l, main="Hubs", vertex.label="",
       3  +    vertex.size=10 * sqrt(hub_score(aidsblog)$vector))
       4  > plot(aidsblog, layout=l, main="Authorities",
       5  +    vertex.label="", vertex.size=10 *
       6  +    sqrt(authority_score(aidsblog)$vector))
```

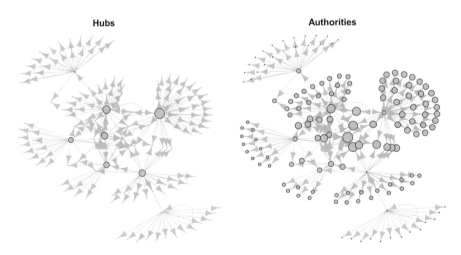

Fig. 4.5 AIDS blog network with vertex area proportional to hubs and authority centrality measures

4.2.3 Characterizing Edges

All of the summary measures discussed so far (i.e., degree and other, more general, notions of centrality) are for vertices, as it seems to be most common in practice that questions of importance are in regard to the vertices of a graph. But some questions are more naturally associated with edges. For example, we might ask which ties in a social network are most important for the spread of, say, information or rumors. Edge betweenness centrality—which extends vertex betweenness centrality in a straightforward manner, by assigning to each edge a value that reflects the number of shortest paths traversing that edge—is a natural quantity to use here.

Using edge betweenness with the karate network and examining, for instance, the edges with the three largest betweenness values

```
#4.11  1 > E(karate)[order(eb, decreasing=T)[1:3]]
       2 + 3/78 edges from 4b458a1 (vertex names):
       3 [1] Actor 20--John A
       4 [2] Mr Hi    --Actor 20
       5 [3] Mr Hi    --Actor 32
```

we are led to note that actor 20 plays a key role from this perspective in facilitating the direct flow of information between the head instructor (Mr Hi, vertex 1) and the administrator (John A, vertex 34).

However, many other vertex centrality measures do not extend as easily. One way around this problem is to apply vertex centrality measures to the vertices in the line graph of a network graph G. The *line graph* of G, say $G' = (V', E')$, is obtained essentially by changing vertices of G to edges, and edges, to vertices, which in **igraph** may be accomplished using make_line_graph. More formally, the vertices $v' \in V'$ represent the original edges $e \in E$, and the edges $e' \in E'$ indicate that the two corresponding original edges in G were incident to a common vertex in G. See Chap. 3 of the edited volume of Brandes and Erlebach [2] for a brief discussion of this approach.

4.3 Characterizing Network Cohesion

A great many questions in network analysis boil down to questions involving *network cohesion*, the extent to which subsets of vertices are cohesive—or 'stuck together'— with respect to the relation defining edges in the network graph. Do friends of a given actor in a social network tend to be friends of one another as well? What collections of proteins in a cell appear to work closely together? Does the structure of the pages in the World Wide Web tend to separate with respect to distinct types of content? What portion of a measured Internet topology would seem to constitute the 'backbone'?

There are many ways that we can define network cohesion, depending on the context of the question being asked. Definitions differ, for example, in scale, ranging from local (e.g., triads) to global (e.g., giant components), and also in the extent to which they are specified explicitly (e.g., cliques) versus implicitly (e.g., 'clusters' or 'communities'). In this section we discuss a number of common ways to define and summarize 'cohesion' in a network graph.

4.3.1 Subgraphs and Censuses

One approach to defining network cohesion is through specification of a certain subgraph(s) of interest. The canonical example of such a subgraph is that of a clique. Recall that cliques are complete subgraphs and hence are subsets of vertices that are fully cohesive, in the sense that all vertices within the subset are connected by edges.

A census of cliques of all sizes can provide some sense of a 'snapshot' of how structured a graph is.

```
#4.12 1 > table(sapply(cliques(karate), length))
      2
      3   1  2  3  4  5
      4  34 78 45 11  2
```

For the karate network a census of this sort reflects that there are 34 nodes (cliques of size one) and 78 edges (cliques of size two), followed by 45 triangles (cliques of size three). The prevalence of the latter are quite evident in the visualizations we have seen (e.g., Fig. 3.5). We also see that the largest cliques are of size five, of which there are only two. These two both involve four actors in common, including actor 1, i.e., the head instructor.

```
#4.13 1 > cliques(karate)[sapply(cliques(karate), length) == 5]
      2 [[1]]
      3 + 5/34 vertices, named, from 4b458a1:
      4 [1] Mr Hi    Actor 2  Actor 3   Actor 4   Actor 14
      5
      6 [[2]]
      7 + 5/34 vertices, named, from 4b458a1:
      8 [1] Mr Hi    Actor 2 Actor 3 Actor 4 Actor 8
```

Note that there is some redundancy in this analysis, in that the cliques of larger sizes necessarily include cliques of smaller sizes. A *maximal* clique is a clique that is not a subset of a larger clique. In the karate network, the two largest cliques (formally called maximum cliques) are maximal, while, for example, the same can be said of only two of the 11 cliques of size four.

```
#4.14 1 > table(sapply(max_cliques(karate), length))
      2
      3   2  3  4  5
      4  11 21  2  2
```

In practice, large cliques are relatively rare, as they necessarily require that a graph G itself be fairly dense, while real-world networks are often sparse. For example, in the

Fig. 4.6 Visual representation of the k-core decomposition of the karate network. Vertices of coreness one (*black*), two (*red*), three (*green*), and four (*blue*) are shown at successively smaller distances from the center, with the same distance for vertices within each core

network of protein–protein interactions in yeast encountered earlier in this chapter, despite being roughly two orders of magnitude larger than the karate network, the size of the largest clique (formally, the *clique number*) is nevertheless comparatively small.

```
#4.15 1 > clique_num(yeast)
      2 [1] 23
```

Various weakened notions of cliques exist. For example, a k-core of a graph G is a subgraph of G for which all vertex degrees are at least k, and such that no other subgraph obeying the same condition contains it (i.e., it is maximal in this property). The notion of cores is particularly popular in visualization, as it provides a way of decomposing a network into 'layers', in the sense of an onion. Such decompositions can be combined in a particularly effective manner with a radial layout (e.g., using a target plot). Figure 4.6 shows the karate network represented in this way.

```
#4.16 1 > cores <- coreness(karate)
      2 > sna::gplot.target(g, cores, circ.lab = FALSE,
      3 +                 circ.col="skyblue", usearrows = FALSE,
      4 +                 vertex.col=cores, edge.col="darkgray")
      5 > detach("package:sna")
      6 > detach("package:network")
```

See [16, Sect. 3.5.2] for an illustration of the same principles in the context of a much larger network, representing a nontrivial portion of the router-level logical topology in the Internet.

Beyond cliques and their variants, there are other classes of subgraphs of common interest in defining network cohesion. Two fundamental quantities, going back to early work in social network analysis [5, 12], are dyads and triads. Dyads are pairs of vertices and, in directed graphs, may take on three possible states: null (no directed edges), asymmetric (one directed edge), or mutual (two directed edges). Similarly, triads are triples of vertices and may take on 16 possible states, ranging from the null subgraph to the subgraph in which all three dyads formed by the vertices in the triad have mutual directed edges.

A census of the possible states of these two classes of subgraphs, i.e., counting how many times each state is observed in a graph G, can yield insight into the nature of the connectivity in the graph. For example, in the AIDS blog network, we see that the vast majority of the dyads are null and, of those that are non-null, almost all are asymmetric, indicating a decided one-sidedness to the manner in which blogs in this network reference each other.

```
#4.17 1 > aidsblog <- simplify(aidsblog)
      2 > dyad_census(aidsblog)
      3 $mut
      4 [1] 3
      5
      6 $asym
      7 [1] 177
      8
      9 $null
     10 [1] 10405
```

This analysis[2] is consistent with the observations from our earlier analysis of hubs and authorities in this network, as shown in Fig. 4.5.

More generally, we note that a census may in principle be conducted of *any* subgraph(s) of interest. Small connected subgraphs of interest are commonly termed *motifs*. The notion of motifs is particularly popular in the study of biological networks, where arguments often are made linking such network substructures to biological function. Examples include subgraphs with a fan-like structure (i.e., multiple directed edges emanating from a single vertex) and feedforward loops (i.e., three directed edges, among three vertices, of the form $\{u, v\}$, $\{v, w\}$, and $\{u, w\}$). In **igraph**, the function `motifs` returns a vector of the number of occurrences of each motif of a prespecified size, while `count_motifs` returns the total number of such motifs.

Note that in large network graphs the enumeration of all occurrences of a given motif can be quite time consuming. For this reason, sampling methods are sometimes used, with motif counts estimated from the sample. The **igraph** function `sample_motifs` employs this approach.

---

[2]Note that we first remove self-loops (of which the original AIDS blog network has three), since the notion of mutuality is well-defined only for dyads.

4.3.2 Density and Related Notions of Relative Frequency

The characterizations of network cohesion described so far proceed by first stating a pre-specified notion of substructure and then looking to see whether it occurs in a graph G and, if so, where and how often. More generally, the related concept of relative frequency can be applied in various useful ways.

The *density* of a graph is the frequency of realized edges relative to potential edges. For example, in a (undirected) graph G with no self-loops and no multiple edges, the density of a subgraph $H = (V_H, E_H)$ is

$$\text{den}(H) = \frac{|E_H|}{|V_H|(|V_H| - 1)/2} \,.$$

(4.4)

The value of $\text{den}(H)$ will lie between zero and one and provides a measure of how close H is to being a clique. In the case that G is a directed graph, the denominator in (4.4) is replaced by $|V_H|(|V_H| - 1)$.

The arguably simple concept of density is made interesting through the freedom we have in the choice of subgraph H defining (4.4). For instance, taking $H = G$ yields the density of the overall graph G. Conversely, taking $H = H_v$ to be the set of neighbors of a vertex $v \in V$, and the edges between them, yields a measure of density in the immediate neighborhood of v.

Applying these ideas to the karate network, for example, we see that the subgraphs corresponding to each of the instructor and the administrator, in union with their immediate respective neighborhoods—i.e., the ego-centric networks around vertices 1 and 34—are noticeably more dense than the overall network.

```
#4.18  1 > ego.instr <- induced_subgraph(karate,
       2 +    neighborhood(karate, 1, 1)[[1]])
       3 > ego.admin <- induced_subgraph(karate,
       4 +    neighborhood(karate, 1, 34)[[1]])
       5 > edge_density(karate)
       6 [1] 0.1390374
       7 > edge_density(ego.instr)
       8 [1] 0.25
       9 > edge_density(ego.admin)
      10 [1] 0.2091503
```

This observation is consistent with the disparity in the number of within-versus between-faction edges in this network, evident in the visualization of Fig. 3.5.

Relative frequency also is used in defining notions of 'clustering' in a graph. For example, the standard use of the term *clustering coefficient* typically refers to the quantity

$$\text{cl}_T(G) = \frac{3\tau_\triangle(G)}{\tau_3(G)} \,,$$

(4.5)

where $\tau_\triangle(G)$ is the number of triangles in the graph G, and $\tau_3(G)$, the number of connected triples (i.e., a subgraph of three vertices connected by two edges, also sometimes called a 2-star). The value $cl_T(G)$ is alternatively called the *transitivity* of the graph, and is a standard quantity of interest in the social network literature, where it is also referred to as the 'fraction of transitive triples.'

Note that $cl_T(G)$ is a measure of global clustering, summarizing the relative frequency with which connected triples close to form triangles. For example, in the karate network we see that only about one quarter of the connected triples close in this manner.

```
#4.19 1 > transitivity(karate)
      2 [1] 0.2556818
```

The local analogue of this measure can also be of interest. Let $\tau_\triangle(v) = \binom{d_v}{2}$ denote the number of triangles in G into which $v \in V$ falls, and $\tau_3(v)$, the number of connected triples in G for which the two edges are both incident to v. The local clustering coefficient is defined as $cl(v) = \tau_\triangle(v)/\tau_3(v)$, for those vertices v with $\tau_3(v) > 0$. In the case of the instructor and administrator of the karate network, for example, we see that their local clustering is only 50–60 % that of the clustering for the network as a whole.

```
#4.20 1 > transitivity(karate, "local", vids=c(1,34))
      2 [1] 0.1500000 0.1102941
```

A concept unique to directed graphs is that of reciprocity, i.e., the extent to which there is reciprocation among ties in a directed network. There are two main approaches to capturing this notion, distinguished by whether the unit of interest in computing relative frequencies is that of dyads or directed edges. In the case that dyads are used as units, reciprocity is defined to be the number of dyads with reciprocated (i.e., mutual) directed edges divided by the number of dyads with a single, unreciprocated edge. Alternatively, reciprocity is defined as the total number of reciprocated edges divided by the total number of edges.

In the AIDS blog network, the reciprocity is quite low by either definition.

```
#4.21 1 > reciprocity(aidsblog, mode="default")
      2 [1] 0.03278689
      3 > reciprocity(aidsblog, mode="ratio")
      4 [1] 0.01666667
```

Recalling our dyad census of the same network, this result is not surprising.

4.3.3 Connectivity, Cuts, and Flows

A basic question of interest is whether a given graph separates into distinct subgraphs. If it does not, we might seek to quantify how close to being able to do so it is. Intimately related to such issues are questions associated with the flow of 'information' in the graph. In this section, we discuss various ways to quantify such notions.

Recall that a graph G is said to be connected if every vertex is reachable from every other (i.e., if for any two vertices, there exists a walk between the two), and that a connected component of a graph is a maximally connected subgraph. Often it is the case that one of the connected components in a graph G dominates the others in magnitude, in that it contains the vast majority of the vertices in G. Such a component is called, borrowing terminology from random graph theory, the *giant component*.

Recall, for example, our network of protein interactions in yeast. The network graph of 2617 vertices is not connected.

```
#4.22 1 > is_connected(yeast)
      2 [1] FALSE
```

A census of all connected components within this graph, however, shows that there clearly is a giant component.

```
#4.23 1 > comps <- decompose(yeast)
      2 > table(sapply(comps, vcount))
      3
      4   2    3    4    5    6    7 2375
      5  63   13    5    6    1    3    1
```

This single component contains $2375/2617 \approx 90\%$ of the vertices in the network. In contrast, none of the other components alone contain even 1 %. In practice, often attention would be restricted to this giant component alone in carrying out further analysis and modeling.

```
#4.24 1 > yeast.gc <- decompose(yeast)[[1]]
```

For example, a celebrated characteristic observed in the giant component of many real-world networks is the so-called *small world* property, which refers to the situation wherein (a) the shortest-path distance between pairs of vertices is generally quite small, but (b) the clustering is relatively high. In our network of protein–protein interactions in yeast, we see that the average path length in the giant component is barely greater than five

```
#4.25 1 > mean_distance(yeast.gc)
      2 [1] 5.09597
```

and even the longest of paths is not much bigger.

```
#4.26 1 > diameter(yeast.gc)
      2 [1] 15
```

Hence, the shortest-path distance in this network scales more like $\log N_v$ rather than N_v, and therefore is considered 'small'. At the same time, the clustering in this network is relatively large

```
#4.27 1 > transitivity(yeast.gc)
      2 [1] 0.4686663
```

indicating that close to 50% of connected triples close to form triangles.

A somewhat more refined notion of connectivity than components derives from asking whether, if an arbitrary subset of k vertices (edges) is removed from a graph, the remaining subgraph is connected. The concepts of vertex- and edge-connectivity, and the related concepts of vertex- and edge-cuts, help to make such notions precise.

A graph G is called *k-vertex-connected* if (i) the number of vertices $N_v > k$, and (ii) the removal of any subset of vertices $X \subset V$ of cardinality $|X| < k$ leaves a subgraph that is connected. Similarly, G is called *k-edge-connected* if (i) $N_v \geq 2$, and (ii) the removal of any subset of edges $Y \subseteq E$ of cardinality $|Y| < k$ leaves a subgraph that is connected. The *vertex (edge) connectivity* of G is the largest integer such that G is k-vertex- (k-edge-) connected. It can be shown that the vertex connectivity is bounded above by the edge connectivity, which in turn is bounded above by the minimum degree d_{min} among vertices in G.

In the case of the giant component of the yeast network, the vertex and edge connectivity are both equal to one.

```
#4.28  1  > vertex_connectivity(yeast.gc)
       2  [1] 1
       3  > edge_connectivity(yeast.gc)
       4  [1] 1
```

Thus it requires the removal of only a single well-chosen vertex or edge in order to break this subgraph into additional components.

If the removal of a particular set of vertices (edges) in a graph disconnects the graph, that set is called a *vertex-cut (edge-cut)*. A single vertex that disconnects the graph is called a *cut vertex*, or sometimes an *articulation point*. Identification of such vertices can provide a sense of where a network is vulnerable (e.g., in the sense of an attack, where disconnecting produces undesired consequences, such as a power outage in an energy network). In the giant component of the yeast network, almost 15% of the vertices are cut vertices.

```
#4.29  1  > yeast.cut.vertices <- articulation_points(yeast.gc)
       2  > length(yeast.cut.vertices)
       3  [1] 350
```

A fundamental result in graph theory, known as Menger's theorem, essentially states that a nontrivial graph G is k-vertex (k-edge) connected if and only if all pairs of distinct vertices $u, v \in V$ can be connected by k vertex-disjoint (edge-disjoint) paths. This result relates the robustness of a graph in the face of removal of its vertices (edges) to the richness of distinct paths running throughout it. A graph with low vertex (edge) connectivity therefore can have the paths, and hence any 'information' flowing over those paths, disrupted by removing an appropriate choice of a correspondingly small number of vertices (edges). Functions like `shortest_paths`, `max_flow`, and `min_cut` in **igraph** can be used to calculate quantities relevant to this connection between cuts and flows.

Most of the concepts introduced above extend to the case of directed graphs in a straightforward manner. For example, recall that a digraph G is weakly connected if its underlying graph is connected, and it is called strongly connected if every vertex v

is reachable from every other vertex u by a directed walk. A strongly connected component is a subgraph that is strongly connected. The concept of vertex-connectivity (edge-connectivity) extends analogously by requiring that the graphs remaining after removal of vertices (edges) be strongly connected. Similarly, the notion of cuts and flows remain essentially unchanged, except that now there is a directionality designated.

Note that the distinction between strong and weak connectivity can be severe for some digraphs. For example, the AIDS blog network is weakly connected

```
#4.30  1 > is_connected(aidsblog, mode=c("weak"))
       2 [1] TRUE
```

but not strongly connected.

```
#4.31  1 > is_connected(aidsblog, mode=c("strong"))
       2 [1] FALSE
```

And while there does exist a strongly connected component within the graph, there is only one and it has only four vertices.

```
#4.32  1 > aidsblog.scc <- components(aidsblog, mode=c("strong"))
       2 > table(aidsblog.scc$csize)
       3
       4     1    4
       5   142    1
```

4.4 Graph Partitioning

Partitioning—broadly speaking—refers to the segmentation of a set of elements into 'natural' subsets. More formally, a *partition* $\mathscr{C} = \{C_1, \ldots, C_K\}$ of a finite set S is a decomposition of S into K disjoint, nonempty subsets C_k such that $\cup_{k=1}^K C_k = S$. In the analysis of network graphs, partitioning is a useful tool for finding, in an unsupervised fashion, subsets of vertices that demonstrate a 'cohesiveness' with respect to the underlying relational patterns.

A 'cohesive' subset of vertices generally is taken to refer to a subset of vertices that (i) are well connected among themselves, and at the same time (ii) are relatively well separated from the remaining vertices. More formally, graph partitioning algorithms typically seek a partition $\mathscr{C} = \{C_1, \ldots, C_K\}$ of the vertex set V of a graph $G = (V, E)$ in such a manner that the sets $E(C_k, C_{k'})$ of edges connecting vertices in C_k to vertices in $C_{k'}$ are relatively small in size compared to the sets $E(C_k) = E(C_k, C_k)$ of edges connecting vertices within the C_k.

This problem of *graph partitioning* is also commonly referred to as *community detection* in the complex networks literature. The development of methods for community detection has been—and continues to be—a highly active area of research. Here we illustrate through the use of two well-established classes of methods—those based on adaptations of hierarchical clustering and those based on spectral partitioning. For an extensive survey of this area, see [17].

4.4.1   Hierarchical Clustering

A great many methods for graph partitioning are essentially variations on the more general concept of *hierarchical clustering* used in data analysis.[3] There are numerous techniques that have been proposed for the general clustering problem, differing primarily in (i) how they evaluate the quality of proposed clusterings and (ii) the algorithms by which they seek to optimize that quality. See Jain, Murty, and Flynn [13], for example. These methods take a greedy approach to searching the space of all possible partitions $\mathscr{C}$, by iteratively modifying successive candidate partitions.

Hierarchical methods are classified as either *agglomerative*, being based on the successive coarsening of partitions through the process of merging, or *divisive*, being based on the successive refinement of partitions through the process of splitting. At each stage, the current candidate partition is modified in a way that minimizes a specified measure of cost. In agglomerative methods, the least costly merge of two previously existing partition elements is executed, whereas in divisive methods, it is the least costly split of a single existing partition element into two that is executed.

The measure of cost incorporated into a hierarchical clustering method used in graph partitioning should reflect our sense of what defines a 'cohesive' subset of vertices. There are many cost measures that have been proposed. A particularly popular measure is that of *modularity* [19]. Let $\mathscr{C} = \{C_1, \ldots, C_K\}$ be a given candidate partition and define $f_{ij} = f_{ij}(\mathscr{C})$ to be the fraction of edges in the original network that connect vertices in C_i with vertices in C_j. The modularity of $\mathscr{C}$ is the value

$$\mathsf{mod}(\mathscr{C}) = \sum_{k=1}^{K} \left[f_{kk}(\mathscr{C}) - f_{kk}^* \right]^2 , \tag{4.6}$$

where f_{kk}^* is the expected value of f_{kk} under some model of random edge assignment. Most commonly, f_{kk}^* is defined to be $f_{k+}f_{+k}$, where f_{k+} and f_{+k} are the kth row and column sums of $\mathbf{f}$, the $K \times K$ matrix[4] formed by the entries f_{ij}. This choice corresponds to a model in which a graph is constructed to have the same degree distribution as G, but with edges otherwise placed at random, without respect to the underlying partition elements dictated by $\mathscr{C}$. Large values of the modularity are therefore taken to suggest that $\mathscr{C}$ captures nontrivial 'group' structure, beyond that expected to occur under the random assignment of edges.

In principle the optimization of the modularity in (4.6) requires a search over all possible partitions $\mathscr{C}$, which is prohibitively expensive in networks of moderate size and larger. A fast, greedy approach to optimization has been proposed in [4], in the form of an agglomerative hierarchical clustering algorithm, and implemented

---

[3]Note that 'clustering' as used here, which is standard terminology in the broader data analysis community, differs from 'clustering' as used in Sect. 4.3.2, which arose in the social network community, in reference to the coefficient cl_T used to summarize the relative density of triangles among connected triples.

[4]Note that for undirected graphs this matrix will be symmetric, and hence $f_{k+} = f_{+k}$; for directed graphs, however, $\mathbf{f}$ can be asymmetric.

in **igraph** as `cluster_fast_greedy`. The result of this and related community detection methods in **igraph** is to produce an object of the class *communities*, which can then serve as input to various other functions.

Applying this method to the karate network,

```
#4.33 1 > kc <- cluster_fast_greedy(karate)
```

we find that the method has declared there to be three communities.

```
#4.34 1 > length(kc)
      2 [1] 3
      3 > sizes(kc)
      4 Community sizes
      5  1  2  3
      6 18 11  5
```

Based on what we know of this network, and examining the community membership

```
#4.35  1 > membership(kc)
       2     Mr Hi   Actor 2   Actor 3   Actor 4   Actor 5   Actor 6
       3         2         2         2         2         3         3
       4   Actor 7   Actor 8   Actor 9  Actor 10  Actor 11  Actor 12
       5         3         2         1         1         3         2
       6  Actor 13  Actor 14  Actor 15  Actor 16  Actor 17  Actor 18
       7         2         2         1         1         3         2
       8  Actor 19  Actor 20  Actor 21  Actor 22  Actor 23  Actor 24
       9         1         2         1         2         1         1
      10  Actor 25  Actor 26  Actor 27  Actor 28  Actor 29  Actor 30
      11         1         1         1         1         1         1
      12  Actor 31  Actor 32  Actor 33    John A
      13         1         1         1         1
```

it would be reasonable to conjecture that the largest community of 18 members is centered around the administrator (i.e., John A, vertex ID 34), while the second largest community of 11 members is centered around the head instructor (i.e., Mr Hi, vertex ID 1). The visual representation obtained by plotting the network with these community designations indicated,

```
#4.36 1 > plot(kc,karate)
```

as seen on the left in Fig. 4.7, provides further support for this conjecture.

Whether agglomerative or divisive, when used for network graph partitioning, hierarchical clustering methods actually produce, as the name indicates, an entire hierarchy of nested partitions of the graph, not just a single partition. These partitions can range fully between the two trivial partitions $\{\{v_1\}, \dots, \{v_{N_v}\}\}$ and V. Agglomerative methods begin with the former of these two, while divisive methods begin with the latter. The resulting hierarchy typically is represented in the form of a tree, called a *dendrogram*.

An example of a dendrogram is shown on the right in Fig. 4.7, for our hierarchical partitioning of the karate network, using the **igraph** function `dendPlot`.

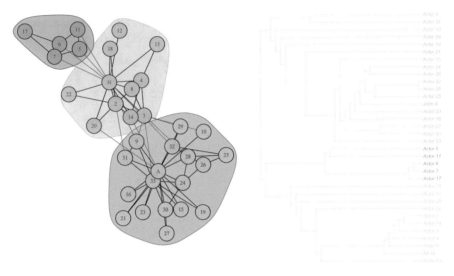

Fig. 4.7 *Left*: Partitioning of the karate network obtained from fast agglomerative hierarchical clustering approximation to optimization of modularity. *Right*: The corresponding dendrogram for this partitioning

```
#4.37  1 > library(ape)
       2 > dendPlot(kc, mode="phylo")
```

The package **ape** is called to facilitate the mode 'phylo' in dendPlot, which uses tools from the former, designed for display of phylogenetic trees, in rendering the dendrogram.

4.4.2   Spectral Partitioning

Another common approach to graph partitioning is to exploit results in spectral graph theory that associate the connectivity of a graph G with the eigen-analysis of certain matrices. Here we look at one popular approach, based on analysis of the so-called Laplacian of a graph. For general background on spectral graph theory, see the monograph by Chung [3].

The *graph Laplacian* of a graph G, with adjacency matrix $\mathbf{A}$, is a matrix $\mathbf{L} = \mathbf{D} - \mathbf{A}$, where $\mathbf{D} = \text{diag}\,[(d_v)]$ is a diagonal matrix with elements $D_{vv} = d_v$ the entries of the degree distribution of G. A formal result in spectral graph theory (e.g., [11, Lemma 13.1.1]) states that a graph G will consist of K connected components if and only if $\lambda_1(\mathbf{L}) = \cdots = \lambda_K(\mathbf{L}) = 0$ and $\lambda_{K+1}(\mathbf{L}) > 0$, where $\lambda_1 \leq \ldots \leq \lambda_{N_v}$ are the (not necessarily distinct) eigenvalues of $\mathbf{L}$, ordered from small to large. Hence, the number of components in a graph is directly related to the number of non-zero eigenvalues of the graph Laplacian.

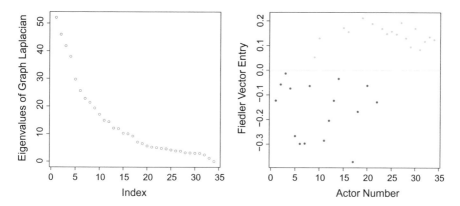

Fig. 4.8 *Left*: Eigenvalues of the graph Laplacian **L** for the karate network. *Right*: The accompanying Fiedler vector and its corresponding partition (indicated by the *gray horizontal line*). The two factions in the club (led by the instructor and the administrator, respectively) are indicated in *red* and *cyan*

The smallest eigenvalue of **L** can be shown to be identically equal to zero, with corresponding eigenvector $\mathbf{x}_1 = (1, \ldots, 1)^T$. Therefore, for example, if we suspect a graph G to consist of 'nearly' $K = 2$ components, and hence to be a good candidate for bisection, we might expect $\lambda_2(\mathbf{L})$ to be close to zero. In fact, such expectations are reasonable, as the value λ_2 is closely related to a number of measures of graph connectivity and structure. In particular, these relationships indicate that the smaller λ_2 the more amenable the graph is to being separated into two subgraphs by severing a relatively small number of edges between the two. See the survey by Mohar [18] or, more recently, the volume edited by Brandes and Erlebach [2, Chap. 14].

Fiedler [6], the first to associate λ_2 with the connectivity of a graph, suggested partitioning vertices by separating them according to the sign of their entries in the corresponding eigenvector $\mathbf{x}_2$. The result is to produce two subsets of vertices (a so-called *cut*)

$$S = \{v \in V : \mathbf{x}_2(v) \geq 0\} \quad \text{and} \quad \bar{S} = \{v \in V : \mathbf{x}_2(v) < 0\} . \tag{4.7}$$

The vector $\mathbf{x}_2$ is hence often called the *Fiedler vector*, and the corresponding eigenvalue λ_2, the *Fiedler value*.

We illustrate using the karate network again. It is straightforward to do the necessary eigen-analysis.

```
#4.38 1 > k.lap <- laplacian_matrix(karate)
      2 > eig.anal <- eigen(k.lap)
```

We plot the eigenvalues of the graph Laplacian in Fig. 4.8

```
#4.39 1 > plot(eig.anal$values, col="blue",
      2 +    ylab="Eigenvalues of Graph Laplacian")
```

and see that (a) there is only one eigenvalue exactly equal to zero (as expected, since this network is connected) and (b) the second smallest eigenvalue λ_2 is quite close to zero. Extracting the Fiedler vector

```
#4.40 1 > f.vec <- eig.anal$vectors[, 33]
```

and plotting the entries of that vector versus actor number,

```
#4.41 1 > faction <- get.vertex.attribute(karate, "Faction")
      2 > f.colors <- as.character(length(faction))
      3 > f.colors[faction == 1] <- "red"
      4 > f.colors[faction == 2] <- "cyan"
      5 > plot(f.vec, pch=16, xlab="Actor Number",
      6 +   ylab="Fiedler Vector Entry", col=f.colors)
      7 > abline(0, 0, lwd=2, col="lightgray")
```

we find that this spectral method exactly captures the partitioning of the network indicated by the faction labels.

In general, of course, we may well expect that a network should be partitioned into more than just two subgraphs. The method of spectral partitioning just described can be applied iteratively, first partitioning a graph into two subgraphs, then each of those two subgraphs into further subgraphs, and so on. Ideally, however, it is desirable that such iterations be aimed at optimizing some common objective function. Newman [7] proposes a method whose technical development parallels that of the spectral bisection method quite closely, but with a matrix related to modularity playing the role of the Laplacian L. This method is implemented in the function `cluster_leading_eigen` in **igraph**.

4.4.3   Validation of Graph Partitioning

Just as in the general problem of clustering data, the question of validation is important for graph partitioning—but often nontrivial. It is generally expected, where cohesive subsets of vertices are present in a network graph, that underlying these subsets there is some commonality in certain relevant characteristics (or attributes) of the vertices. For example, proteins found to cluster in a graph of protein interactions often participate in similar biological processes; similarly, actors found to cluster in a social network may share certain interests.

Graph partitioning may be viewed as a tool for discovering such subsets in the absence of knowledge of these characteristics. When we do have knowledge of some externally defined notion of class membership, it can be interesting to compare and contrast the resulting assignments with those deriving from graph partitioning. As an illustration, consider our network of protein–protein interactions in yeast. These particular yeast data include the assignment of proteins to one of 13 functional classes (including "unknown", denote by 'U').

```
#4.42 1 > func.class <- vertex_attr(yeast.gc, "Class")
      2 > table(func.class)
```

```
3 func.class
4   A   B   C   D   E   F   G   M   O   P   R   T   U
5  51  98 122 238  95 171  96 278 171 248  45 240 483
```

These classes are a way of categorizing the roles of proteins in helping the cell accomplish various tasks through higher-level cellular processes.

The affinity of proteins to physically bind to each other is known to be directly related to their participation in common cellular functions. Hence, the external assignment of proteins to functional classes should correlate, to at least some extent, with their assignment to 'communities' by a reasonable graph partitioning algorithm. Using the same hierarchical clustering algorithm as in the previous section, now applied to the (giant component of the) yeast network,

```
#4.43 1 > yc <- cluster_fast_greedy(yeast.gc)
      2 > c.m <- membership(yc)
```

a simple two-dimensional table allows us to group proteins according to their membership under each categorization.

```
#4.44 1 > table(c.m, func.class, useNA=c("no"))
```

		func.class												
c.m	A	B	C	D	E	F	G	M	O	P	R	T	U
1	0	0	0	1	3	7	0	6	3	110	2	35	14
2	0	2	2	7	1	1	1	4	39	5	0	4	27
3	1	9	7	18	4	8	4	20	10	23	8	74	64
4	25	11	10	22	72	84	81	168	14	75	16	27	121
5	1	7	5	14	0	4	0	2	3	6	1	34	68
6	1	24	1	4	1	4	0	7	0	1	0	19	16
7	6	18	6	76	7	9	3	7	8	5	1	7	33
8	8	12	67	59	1	34	0	19	60	10	7	6	73
9	4	1	7	7	2	10	5	3	2	0	3	0	11
10	0	0	0	6	0	0	0	2	0	5	0	11	1
11	0	9	0	10	1	3	0	0	0	0	0	2	4
12	0	1	3	0	0	0	0	6	10	0	0	0	2
13	0	1	1	2	0	1	0	0	2	0	0	16	10
14	1	0	4	1	0	1	0	0	4	0	1	0	11
15	0	1	0	0	0	2	0	2	0	0	1	0	8
16	0	1	2	0	0	1	0	0	10	0	0	0	0
17	0	0	1	3	0	0	0	2	0	0	0	2	3
18	0	0	0	0	3	1	0	9	0	0	1	0	1
19	0	1	1	1	0	0	0	0	0	0	0	0	3
20	0	0	0	6	0	0	0	1	0	0	0	1	2
21	1	0	0	0	0	0	0	0	6	0	0	1	0
22	0	0	0	0	0	0	0	1	0	0	0	0	8
23	0	0	0	0	0	0	0	4	0	0	0	0	0
24	0	0	0	0	0	0	2	2	0	0	0	1	0
25	0	0	0	0	0	0	0	5	0	0	0	0	0
26	0	0	1	0	0	0	0	4	0	0	1	0	1
27	3	0	4	0	0	1	0	0	0	0	0	0	0
28	0	0	0	0	0	0	0	0	0	6	0	0	0
29	0	0	0	1	0	0	0	1	0	0	3	0	0
30	0	0	0	0	0	0	0	0	0	2	0	0	2
31	0	0	0	0	0	0	0	3	0	0	0	0	0

Some of the membership assignments resulting from our algorithm overlap quite strongly with individual functional classes. For example, 110 of the 182 proteins in the first community have the functional class "P" (indicating a role in protein synthesis), which suggests that the first community is largely capturing that class. On the other hand, the 733 proteins in the fourth community are spread through all functional classes (including 121 with unknown function), making this community decidedly less interpretable against this validation set.

4.5   Assortativity and Mixing

Selective linking among vertices, according to a certain characteristic(s), is termed *assortative mixing* in the social network literature. Measures that quantify the extent of assortative mixing in a given network have been referred to as *assortativity coefficients*, and are essentially variations on the concept of correlation coefficients.

Note that the vertex characteristics involved can be categorical, ordinal, or continuous. Consider the categorical case, and suppose that each vertex in a graph G can be labeled according to one of M categories. The assortativity coefficient in this setting is defined to be

$$r_a = \frac{\sum_i f_{ii} - \sum_i f_{i+} f_{+i}}{1 - \sum_i f_{i+} f_{+i}} \ , \tag{4.8}$$

where f_{ij} is the fraction of edges in G that join a vertex in the ith category with a vertex in the jth category, and f_{i+} and f_{+i} denote the ith marginal row and column sums, respectively, of the resulting matrix[5] $\mathbf{f}$.

The value r_a lies between -1 and 1. It is equal to zero when the mixing in the graph is no different from that obtained through a random assignment of edges that preserves the marginal degree distribution (see the related discussion surrounding (4.6) above). Similarly, it is equal to one when there is perfect assortative mixing (i.e., when edges only connect vertices of the same category). However, in the event that the mixing is perfectly disassortative, in the sense that every edge in the graph connects vertices of two different categories, the coefficient in (4.8) need not take the value -1. See [8] for discussion.

Consider again our network of protein–protein interactions in yeast. The fact that physical binding of proteins is known to be directly relevant to functional classes suggests that there will frequently be strong assortative mixing in protein–protein interaction networks with respective to these classes as attributes. For instance, for the class 'P', representing proteins that are known to play a role in protein synthesis, we see an assortativity coefficient of over 0.5.

```
#4.45  1  > v.types <- (V(yeast)$Class=="P")+1
       2  > v.types[is.na(v.types)] <- 1
       3  > assortativity_nominal(yeast, v.types, directed=FALSE)
       4  [1] 0.5232879
```

---

[5]Note that these quantities are defined in complete analogy to the same quantities that underlie our definition of modularity.

When the vertex characteristic of interest is continuous, rather than discrete, denote by (x_e, y_e) the values of that characteristic for the vertices joined by an edge $e \in E$. Then a natural candidate for quantifying the assortativity in this characteristic is just the Pearson correlation coefficient of the pairs (x_e, y_e),

$$r = \frac{\sum_{x,y} xy \left(f_{xy} - f_{x+} f_{+y} \right)}{\sigma_x \sigma_y} . \tag{4.9}$$

Here the summation is over all unique observed pairs (x, y), with f_{xy}, f_{x+}, and f_{+y} defined in analogy to the categorical case, and σ_x and σ_y are the standard deviations corresponding to the distribution of frequencies $\{f_{x+}\}$ and $\{f_{+y}\}$, respectively.

One common use of the assortativity coefficient r in (4.9) is in summarizing degree–degree correlation of adjacent vertices, when studying network structure. For the yeast network, we find that the degree correlation is positive and relatively large.

```
#4.46 1 > assortativity_degree(yeast)
      2 [1] 0.4610798
```

This observation is consistent with the analysis in Fig. 4.3.

4.6 Additional Reading

The array of tools and techniques available for the description and summary of structural characteristics of a network graph is quite large and still growing. Here in this chapter we have contented ourselves simply with presenting material on topics arguably at the core of this area. The volume edited by Brandes and Erlebach [2] provides a much more detailed—yet highly readable—presentation of most of the topics in this chapter, as well a number of others not covered here, from the perspective of computer science and applied graph theory. The book by Newman [20] presents a similarly detailed treatment from the perspective of statistical physics, as does the book by Fornito, Zalesky, and Bullmore [9], from the perspective of computational neuroscience. Finally, in the literature on social networks, two canonical references are the small volume by Scott [22] and the much larger volume by Wasserman and Faust [23].

References

1. P. Bonacich, "Factoring and weighting approaches to status scores and clique identification," *Journal of Mathematical Sociology*, vol. 2, no. 1, pp. 113–120, 1972.
2. U. Brandes and T. Erlebach, Eds., *Network Analysis: Methodological Foundations*, ser. Lecture Notes in Computer Science. New York: Springer-Verlag, 2005, vol. 3418.

3. F. Chung, *Spectral Graph Theory*. American Mathematical Society, 1997.
4. A. Clauset, M. Newman, and C. Moore, "Finding community structure in very large networks," *Physical Review E*, vol. 70, no. 6, p. 66111, 2004.
5. J. Davis and S. Leinhardt, "The structure of positive interpersonal relations in small groups," in *Sociological Theories in Progress*, J. Berger, Ed. Boston: Houghton-Mifflin, 1972, vol. 2.
6. M. Fiedler, "Algebraic connectivity of graphs," *Czechoslovak Mathematical Journal*, vol. 23, no. 98, pp. 298–305, 1973.
7. M. Fiedler, "Finding community structure in networks using the eigenvectors of matrices," *Physical Review E*, vol. 74, no. 3, p. 36104, 2006.
8. M. Fiedler, "Mixing patterns in networks," *Physical Review E*, vol. 67, no. 2, p. 26126, 2003.
9. A. Fornito, A. Zalesky, and E. Bullmore, *Fundamentals of Brain Network Analysis*. Academic Press, 2016.
10. L. Freeman, "A set of measures of centrality based on betweenness," *Sociometry*, vol. 40, no. 1, pp. 35–41, 1977.
11. C. Godsil and G. Royle, *Algebraic Graph Theory*. New York: Springer, 2001.
12. P. Holland and S. Leinhardt, "A method for detecting structure in sociometric data," *American Journal of Sociology*, vol. 70, pp. 492 – 513, 1970.
13. A. Jain, M. Murty, and P. Flynn, "Data clustering: a review," *ACM Computing Surveys*, vol. 31, no. 3, pp. 264–323, 1999.
14. L. Katz, "A new status index derived from sociometric analysis," *Psychometrika*, vol. 18, no. 1, pp. 39–43, 1953.
15. J. Kleinberg, "Authoritative sources in a hyperlinked environment," *Journal of the ACM*, vol. 46, no. 5, pp. 604–632, 1999.
16. E. Kolaczyk, *Statistical Analysis of Network Data: Methods and Models*. Springer Verlag, 2009.
17. A. Lancichinetti and S. Fortunato, "Community detection algorithms: A comparative analysis," *Physical review E*, vol. 80, no. 5, p. 056117, 2009.
18. B. Mohar, "The Laplacian spectrum of graphs," in *Graph Theory, Combinatorics, and Applications*, Y. Alavi, G. Chartrand, O. Oellermann, and A. Schwenk, Eds. New York: Wiley, 1991, vol. 2, pp. 871–898.
19. M. Newman and M. Girvan, "Finding and evaluating community structure in networks," *Physical Review E*, vol. 69, no. 2, p. 26113, 2004.
20. M. Newman, *Networks: an Introduction*. Oxford University Press, Inc., 2010.
21. G. Sabidussi, "The centrality index of a graph," *Psychometrika*, vol. 31, pp. 581–683, 1966.
22. J. Scott, *Social Network Analysis: A Handbook, Second Edition*. Newbury Park, CA: Sage Publications, 2004.
23. S. Wasserman and K. Faust, *Social Network Analysis: Methods and Applications*. New York: Cambridge University Press, 1994.

Chapter 5
Mathematical Models for Network Graphs

5.1 Introduction

So far in this book, the emphasis has been almost entirely focused upon methods, to the exclusion of modeling—methods for constructing network graphs, for visualizing network graphs, and for characterizing their observed structure. For the remainder of this book, our focus will shift to the construction and use of models in the analysis of network data, beginning with this chapter, in which we turn to the topic of modeling network graphs.

By a model for a network graph we mean effectively a collection

$$\{ \mathbb{P}_\theta(G), G \in \mathscr{G} : \theta \in \Theta \} , \tag{5.1}$$

where $\mathscr{G}$ is a collection (or 'ensemble') of possible graphs, $\mathbb{P}_\theta$ is a probability distribution on $\mathscr{G}$, and θ is a vector of parameters, ranging over possible values in Θ. When convenient, we may often drop the explicit reference to θ and simply write $\mathbb{P}$ for the probability function.

In practice, network graph models are used for a variety of purposes. These include the testing for 'significance' of a pre-defined characteristic(s) in a given network graph, the study of proposed mechanisms for generating certain commonly observed properties in real-world networks (such as broad degree distributions or small-world effects), or the assessment of potential predictive factors of relational ties.

The richness of network graph modeling derives largely from how we choose to specify $\mathbb{P}(\cdot)$, with methods in the literature ranging from the simple to the complex. It is useful for our purposes to distinguish, broadly speaking, between models defined more from (i) a mathematical perspective, versus (ii) a statistical perspective. Those of the former class tend to be simpler in nature and more amendable to mathematical analysis yet, at the same time, do not always necessarily lend themselves well to formal statistical techniques of model fitting and assessment. On the other hand, those of the latter class typically are designed to be fit to data, but their mathematical

© Springer Nature Switzerland AG 2020
E. D. Kolaczyk and G. Csárdi, *Statistical Analysis of Network Data with R*, Use R!,
https://doi.org/10.1007/978-3-030-44129-6_5

analysis can be challenging in some cases. Nonetheless, both classes of network graph models have their uses for analyzing network graph data.

We will examine certain mathematical models for network graphs in this chapter, and various statistical models, in the following chapter. We consider classical random graph models in Sect. 5.2, and more recent generalizations, in Sect. 5.3. In both cases, these models dictate simply that a graph be drawn uniformly at random from a collection $\mathscr{G}$, but differ in their specification of $\mathscr{G}$. Other classes of network graph models are then presented in Sect. 5.4, where the focus is on models designed to mimic certain observed 'real-world' properties, often through the incorporation of an explicit mechanism(s). Finally, in Sect. 5.5, we look at several examples of ways in which these various mathematical models for network graphs can be used for statistical purposes.

5.2 Classical Random Graph Models

The term *random graph model* typically is used to refer to a model specifying a collection $\mathscr{G}$ and a uniform probability $\mathbb{P}(\cdot)$ over $\mathscr{G}$. Random graph models are arguably the most well-developed class of network graph models, mathematically speaking.

The classical theory of random graph models, as established in a series of seminal papers by Erdős and Rényi [1–3], rests upon a simple model that places equal probability on all graphs of a given order and size. Specifically, their model specifies a collection $\mathscr{G}_{N_v,N_e}$ of all graphs $G = (V, E)$ with $|V| = N_v$ and $|E| = N_e$, and assigns probability $\mathbb{P}(G) = \binom{N}{N_e}^{-1}$ to each $G \in \mathscr{G}_{N_v,N_e}$, where $N = \binom{N_v}{2}$ is the total number of distinct vertex pairs.

A variant of $\mathscr{G}_{N_v,N_e}$, first suggested by Gilbert [4] at approximately the same time, arguably is seen more often in practice. In this formulation, a collection $\mathscr{G}_{N_v,p}$ is defined to consist of all graphs G of order N_v that may be obtained by assigning an edge independently to each pair of distinct vertices with probability $p \in (0, 1)$. Accordingly, this type of model sometimes is referred to as a Bernoulli random graph model. When p is an appropriately defined function of N_v, and $N_e \sim pN_v^2$, these two classes of models are essentially equivalent for large N_v.

The functions `sample_gnm` and `sample_gnp` in **igraph** can be used to simulate classical random graphs of either type. Figure 5.1 shows a realization of a classical random graph, based on the choice of $N_v = 100$ vertices and a probability of $p = 0.02$ of an edge between any pair of vertices. Using a circular layout, we can see that the edges appear to be scattered between vertex pairs in a fairly uniform manner, as expected.

```
#5.1 1 > library(sand)
     2 > set.seed(42)
     3 > g.er <- sample_gnp(100, 0.02)
     4 > plot(g.er, layout=layout_in_circle, vertex.label=NA)
```

Note that random graphs generated in the manner described above need not be connected.

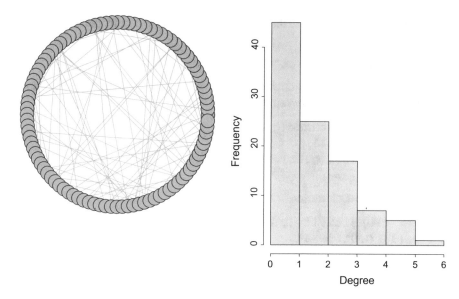

Fig. 5.1 *Left*: A random graph simulated according to an Erdős-Rényi model. *Right*: The corresponding degree distribution

```
#5.2 1 > is_connected(g.er)
     2 [1] FALSE
```

Although this particular realization is not connected, it does nevertheless have a giant component, containing 71 of the 100 vertices. All other components contain between one and four vertices only.

```
#5.3 1 > table(sapply(decompose(g.er), vcount))
     2
     3  1  2  3  4 71
     4 15  2  2  1  1
```

In general, a classical random graph G will with high probability have a giant component if $p = c/N_v$ for some $c > 1$.

Under this same parameterization for p, for $c > 0$, the degree distribution will be well-approximated by a Poisson distribution, with mean c, for large N_v. That this should be true is somewhat easy to see at an intuitive level, since the degree of any given vertex is distributed as a binomial random variable, with parameters $N_v - 1$ and p. The formal proof of this result, however, is nontrivial. The book by Bollobás [5] is a standard reference for this and other such results.

Indeed, in our simulated random graph, the mean degree is quite close to the expected value of $(100 - 1) \times 0.02 = 1.98$.

```
#5.4 1 > mean(degree(g.er))
     2 [1] 1.9
```

Furthermore, in Fig. 5.1 we see that the degree distribution is quite homogeneous.

```
#5.5 1 > hist(degree(g.er), col="lightblue",
     2 +    xlab="Degree", ylab="Frequency", main="")
```

Other properties of classical random graphs include that there are relatively few vertices on shortest paths between vertex pairs

```
#5.6 1 > mean_distance(g.er)
     2 [1] 5.276511
     3 > diameter(g.er)
     4 [1] 14
```

and that there is low clustering.

```
#5.7 1 > transitivity(g.er)
     2 [1] 0.01639344
```

More specifically, under the conditions above, it can be shown that the diameter varies like $O(\log N_v)$, and the clustering coefficient, like N_v^{-1}.

5.3   Generalized Random Graph Models

The formulation of Erdős and Rényi can be generalized in a straightforward manner. Specifically, the basic recipe is to (a) define a collection of graphs $\mathscr{G}$ consisting of all graphs of a fixed order N_v that possess a given characteristic(s), and then (b) assign equal probability to each of the graphs $G \in \mathscr{G}$. In the Erdős-Rényi model, for example, the common characteristic is simply that the size of the graphs G be equal to some fixed N_e.

Beyond Erdős-Rényi, the most commonly chosen characteristic is that of a fixed degree sequence. That is, $\mathscr{G}$ is defined to be the collection of all graphs G with a pre-specified degree sequence, which we will write here as $\{d_{(1)}, \ldots, d_{(N_v)}\}$, in ordered form.

The **igraph** function sample_degseq can be used to uniformly sample random graphs with fixed degree sequence. Suppose, for example, that we are interested in graphs of $N_v = 8$ vertices, half of which have degree $d = 2$, and the other half, degree $d = 3$. Two examples of such graphs, drawn uniformly from the collection of all such graphs, are shown in Fig. 5.2.

```
#5.8 1 > degs <- c(2,2,2,2,3,3,3,3)
     2 > g1 <- sample_degseq(degs, method="vl")
     3 > g2 <- sample_degseq(degs, method="vl")
     4 > plot(g1, vertex.label=NA)
     5 > plot(g2, vertex.label=NA)
```

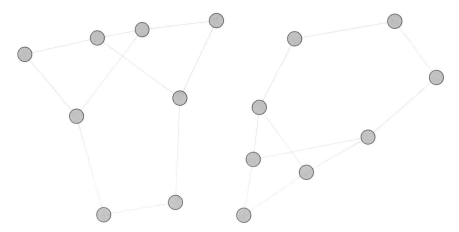

Fig. 5.2 Two graphs drawn uniformly from the collection of all graphs G with $N_v = 8$ vertices and degree sequence $(2, 2, 2, 2, 3, 3, 3, 3)$

Note that these two graphs do indeed differ, in that they are not isomorphic.[1]

```
#5.9  1  > isomorphic(g1, g2)
      2  [1] FALSE
```

For a fixed number of vertices N_v, the collection of random graphs with fixed degree sequence all have the same number of edges N_e, due to the relation $\bar{d} = 2N_e/N_v$, where $\bar{d}$ is the mean degree of the sequence $(d_{(1)}, \ldots, d_{(N_v)})$.

```
#5.10  1  > c(ecount(g1), ecount(g2))
       2  [1] 10 10
```

Therefore, this collection is strictly contained within the collection of random graphs $\mathscr{G}_{N_v, N_e}$, with the corresponding number N_v, N_e of vertices and edges. So the addition of an assumed form for the degree sequence is in this case equivalent to specifying our model through a conditional distribution on the original collection $\mathscr{G}_{N_v, N_e}$.

On the other hand, it is important to keep in mind that all other characteristics are free to vary to the extent allowed by the chosen degree sequence. For example, we can generate a graph with the same degree sequence as our network of protein–protein interactions in yeast.

```
#5.11  1  > data(yeast)
       2  > degs <- degree(yeast)
       3  > fake.yeast <- sample_degseq(degs, method=c("vl"))
       4  > all(degree(yeast) == degree(fake.yeast))
       5  [1] TRUE
```

But the original network has nearly twice the diameter of the simulated version

---

[1] Recall that two graphs are said to be *isomorphic* if they differ only up to relabellings of their vertices and edges that leave the structure unchanged.

```
#5.12  1 > diameter(yeast)
       2 [1] 15
       3 > diameter(fake.yeast)
       4 [1] 8
```

and virtually all of the substantial amount of clustering originally present is now gone.

```
#5.13  1 > transitivity(yeast)
       2 [1] 0.4686178
       3 > transitivity(fake.yeast)
       4 [1] 0.04026804
```

In principle, it is easy to further constrain the definition of the class $\mathscr{G}$, so that additional characteristics beyond the degree sequence are fixed. Markov chain Monte Carlo (MCMC) methods are popular for generating generalized random graphs G from such collections, where the states visited by the Markov chain are the distinct graphs G themselves. Snijders [6], Roberts [7], and more recently, McDonald, Smith, and Forster [8], for example, offer MCMC algorithms for a number of different cases, such as where the graphs G are directed and constrained to have both a fixed pair of in- and out-degree sequences and a fixed number of mutual arcs. However, the design of such algorithms in general is nontrivial, and development of the corresponding theory, in turn, which would allow users to verify the assumptions underlying Markov chain convergence, has tended to lag somewhat behind the pace of algorithm development.

5.4 Network Graph Models Based on Mechanisms

Arguably one of the more important innovations in modern network graph modeling is a movement from traditional random graph models, like those in the previous two sections, to models explicitly designed to mimic certain observed 'real-world' properties, often through the incorporation of a simple mechanism(s). Here in this section we look at two canonical classes of such network graph models.

5.4.1 Small-World Models

Work on modeling of this type received a good deal of its impetus from the seminal paper of Watts and Strogatz [9] and the 'small-world' network model introduced therein. These authors were intrigued by the fact that many networks in the real world display high levels of clustering, but small distances between most nodes. Such behavior cannot be reproduced, for example, by classical random graphs, since, recall that, while the diameter scales like $O(\log N_v)$, indicating small distances between nodes, the clustering coefficient behaves like N_v^{-1}, which suggests very little clustering.

Fig. 5.3 Example of a
Watts-Strogatz 'small-world'
network graph

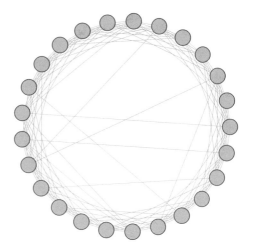

In order to create a network graph with both of these properties, Watts and Strogatz
suggested instead beginning with a graph with lattice structure, and then randomly
'rewiring' a small percentage of the edges. More specifically, in this model we begin
with a set of N_v vertices, arranged in a periodic fashion, and join each vertex to r of
its neighbors to each side. Then, for each edge, independently and with probability
p, one end of that edge will be moved to be incident to another vertex, where that
new vertex is chosen uniformly, but with attention to avoid the construction of loops
and multi-edges.

An example of a small-world network graph of this sort can be generated in **igraph**
using the function sample_smallworld.

```
#5.14  1 > g.ws <- sample_smallworld(1, 25, 5, 0.05)
       2 > plot(g.ws, layout=layout_in_circle, vertex.label=NA)
```

The resulting graph, with $N_v = 25$ vertices, neighborhoods of size $r = 5$, and
rewiring probability $p = 0.05$, is shown in Fig. 5.3.

For the lattice alone, which we generate by setting $p = 0$, there is a substantial
amount of clustering.

```
#5.15  1 > g.lat100 <- sample_smallworld(1, 100, 5, 0)
       2 > transitivity(g.lat100)
       3 [1] 0.6666667
```

But the distance between vertices is non-trivial.

```
#5.16  1 > diameter(g.lat100)
       2 [1] 10
       3 > mean_distance(g.lat100)
       4 [1] 5.454545
```

The effect of rewiring a relatively small number of edges in a random fashion is to
noticeably reduce the distance between vertices, while still maintaining a similarly
high level of clustering.

```
#5.17  1 > g.ws100 <- sample_smallworld(1, 100, 5, 0.05)
       2 > diameter(g.ws100)
       3 [1] 5
       4 > mean_distance(g.ws100)
       5 [1] 2.748687
       6 > transitivity(g.ws100)
       7 [1] 0.5166263
```

This effect may be achieved even with very small p. To illustrate, we simulate according to a particular Watts-Strogatz small-world network model, with $N_v = 1,000$ and $r = 10$, and re-wiring probability p, as p varies over a broad range.

```
#5.18  1 > steps <- seq(-4, -0.5, 0.1)
       2 > len <- length(steps)
       3 > cl <- numeric(len)
       4 > apl <- numeric(len)
       5 > ntrials <- 100
       6 > for (i in (1:len)) {
       7 +    cltemp <- numeric(ntrials)
       8 +    apltemp <- numeric(ntrials)
       9 +    for (j in (1:ntrials)) {
      10 +      g <- sample_smallworld(1, 1000, 10, 10^steps[i])
      11 +      cltemp[j] <- transitivity(g)
      12 +      apltemp[j] <- mean_distance(g)
      13 +    }
      14 +    cl[i] <- mean(cltemp)
      15 +    apl[i] <- mean(apltemp)
      16 + }
```

The results shown in Fig. 5.4, where approximate expected values for normalized versions of average path length and clustering coefficient are plotted, indicate that over a substantial range of p the network exhibits small average distance while maintaining a high level of clustering.

```
#5.19  1 > plot(steps, cl/max(cl), ylim=c(0, 1), lwd=3, type="l",
       2 +    col="blue", xlab=expression(log[10](p)),
       3 +    ylab="Clustering and Average Path Length")
       4 > lines(steps, apl/max(apl), lwd=3, col="red")
```

5.4.2   Preferential Attachment Models

Many networks grow or otherwise evolve in time. The World Wide Web and scientific citation networks are two obvious examples. Similarly, many biological networks may be viewed as evolving as well, over appropriately defined time scales. Much energy has been invested in the development of models that mimic network growth.

In this arena, typically a simple mechanism(s) is specified for how the network changes at any given point in time, based on concepts like vertex preference, fitness, copying, age, and the like. A celebrated example of such a mechanism is that of *preferential attachment*, designed to embody the principle that 'the rich get richer.'

Fig. 5.4 Plot of the clustering coefficient cl(G) (*blue*) and average path length (*red*), as a function of the rewiring probability p for a Watts-Strogatz small-world model. Results are averages based on 100 simulation trials

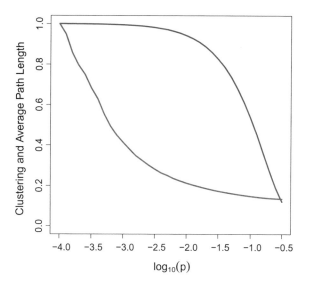

A driving motivation behind the introduction of this particular mechanism was a desire to reproduce the types of broad degree distributions observed in many large, real-world networks. Although there were a number of precursors with similar ideas, it is the work of Barabási and Albert [10] that launched the present-day fascination with models of this type.[2]

Barabási and Albert were motivated by the growth of the World Wide Web, noting that often web pages to which many other pages point will tend to accumulate increasingly greater numbers of links as time goes on. The Barabási-Albert (BA) model for undirected graphs is formulated as follows. Start with an initial graph $G^{(0)}$ of $N_v^{(0)}$ vertices and $N_e^{(0)}$ edges. Then, at stage $t = 1, 2, \ldots$, the current graph $G^{(t-1)}$ is modified to create a new graph $G^{(t)}$ by adding a new vertex of degree $m \geq 1$, where the m new edges are attached to m different vertices in $G^{(t-1)}$, and the probability that the new vertex will be connected to a given vertex v is given by

$$\frac{d_v}{\sum_{v' \in V} d_{v'}} \, . \tag{5.2}$$

That is, at each stage, m existing vertices are connected to a new vertex in a manner preferential to those with higher degrees. After t iterations, the resulting graph $G^{(t)}$ will have $N_v^{(t)} = N_v^{(0)} + t$ vertices and $N_e^{(t)} = N_e^{(0)} + tm$ edges. And, because of the tendency towards preferential attachment, intuitively we would expect that a number of vertices of comparatively high degree should gradually emerge as t increases.

Using the **igraph** function `sample_pa`, we can simulate a BA random graph of, for example, $N_v = 100$ vertices, with $m = 1$ new edges added for each new vertex.

---

[2]For an extensive history of power-law models, including their use in network modeling, see Mitzenmacher [11].

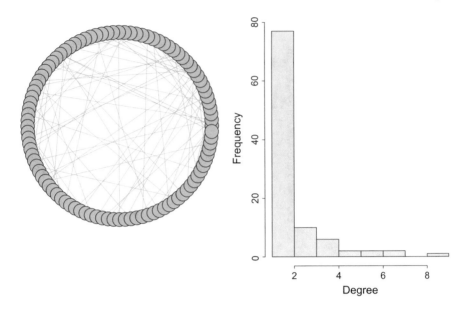

Fig. 5.5 *Left*: A random graph simulated according to a Barabási-Albert model. *Right*: The corresponding degree distribution

```
#5.20 1 > set.seed(42)
      2 > g.ba <- sample_pa(100, directed=FALSE)
```

A visualization of this graph is shown in Fig. 5.5.

```
#5.21 1 > plot(g.ba, layout=layout_in_circle, vertex.label=NA)
```

Note that the edges are spread among vertex pairs in a decidedly less uniform manner than in the classical random graph we saw in Fig. 5.1. And, in fact, there appear to be vertices of especially high degree—so-called 'hub' vertices.

Examination of the degree distribution (also shown in Fig. 5.5)

```
#5.22 1 > hist(degree(g.ba), col="lightblue",
      2 +    xlab="Degree", ylab="Frequency", main="")
```

confirms this suspicion, and indicates, moreover, that the overall distribution is quite heterogeneous. Actually, the vast majority of vertices have degree no more than two in this graph, while, on the other hand, one vertex has a degree of 9.

```
#5.23 1 > summary(degree(g.ba))
      2    Min. 1st Qu.  Median    Mean 3rd Qu.    Max.
      3    1.00    1.00    1.00    1.98    2.00    9.00
```

The most celebrated property of such preferential attachment models is that, in the limit as t tends to infinity, the graphs $G^{(t)}$ have degree distributions that tend to a power-law form $d^{-\alpha}$, with $\alpha = 3$. See [10, 12–14]. This behavior is in noted contrast to the case of classical random graphs.

On the other hand, network graphs generated according to the BA model will share with their classical counterparts the tendency towards relatively few vertices on shortest paths between vertex pairs

```
#5.24 1 > mean_distance(g.ba)
      2 [1] 5.815556
      3 > diameter(g.ba)
      4 [1] 12
```

and low clustering.

```
#5.25 1 > transitivity(g.ba)
      2 [1] 0
```

See [15, 16] for details.

5.5 Assessing Significance of Network Graph Characteristics

As we remarked at the start of this chapter, the network graph models we have described above generally are too simple for serious statistical modeling with observed networks. Nonetheless, from a statistical hypothesis testing perspective, they still have a useful role to play in network analysis. Specifically, the network models introduced in this chapter are frequently used in the assessment of significance of network graph characteristics.

Suppose that we have a graph derived from observations of some sort, which we will denote as G^{obs} here, and that we are interested in some structural characteristic, say $\eta(\cdot)$. In many contexts it is natural to ask whether $\eta(G^{obs})$ is 'significant,' in the sense of being somehow unusual or unexpected. Network models of the type we have considered so far are used in establishing a well-defined frame of reference. That is, for a given collection of graphs $\mathscr{G}$, the value $\eta(G^{obs})$ is compared to the collection of values $\{\eta(G) : G \in \mathscr{G}\}$. If $\eta(G^{obs})$ is judged to be extreme with respect to this collection, then that is taken as evidence that G^{obs} is unusual in having this value.

When random graph models are used, it is straightforward to create a formal reference distribution which, under the accompanying assumption of uniform probability of elements in $\mathscr{G}$, takes the form

$$\mathbb{P}_{\eta,\mathscr{G}}(t) = \frac{\#\{G \in \mathscr{G} : \eta(G) \leq t\}}{|\mathscr{G}|} . \qquad (5.3)$$

If $\eta(G^{obs})$ is found to be sufficiently unlikely under this distribution, this is taken as evidence against the hypothesis that G^{obs} is a uniform draw from $\mathscr{G}$.

This principle lies at the heart of methods aimed at the detection of *network motifs*, defined by Alon and colleagues [17, 18] to be small subgraphs occurring far more frequently in a given network than in comparable random graphs. We refer the reader to these articles for details. Also see [19, Sect. 6.2].

Here we demonstrate similar use of this principle through two somewhat simpler examples.

5.5.1 Assessing the Number of Communities in a Network

Recall from Sect. 4.4.1 that our use of hierarchical clustering, through the function `cluster_fast_greedy`, resulted in the discovery of three communities in the karate network. We might ask ourselves whether this is in some sense unexpected or unusual. In defining our frame of reference we will use two choices of $\mathscr{G}$: (i) graphs of the same order $N_v = 34$ and size $N_e = 78$ as the karate network, and (ii) graphs that obey the further restriction that they have the same degree distribution as the original.

Monte Carlo methods, using some of the same R functions encountered above, allow us to quickly generate approximations to the corresponding reference distributions. In order to do so, we need the order, size, and degree sequence of the original karate network.

```
#5.26 1 > data(karate)
      2 > nv <- vcount(karate)
      3 > ne <- ecount(karate)
      4 > degs <- degree(karate)
```

Over 1000 trials,

```
#5.27 1 > ntrials <- 1000
```

we then generate classical random graphs of this same order and size and, for each one, we use the same community detection algorithm to determine the number of communities.

```
#5.28 1 > num.comm.rg <- numeric(ntrials)
      2 > for(i in (1:ntrials)){
      3 +    g.rg <- sample_gnm(nv, ne)
      4 +    c.rg <- cluster_fast_greedy(g.rg)
      5 +    num.comm.rg[i] <- length(c.rg)
      6 + }
```

Similarly, we do the same using generalized random graphs constrained to have the required degree sequence.

Fig. 5.6 Distribution of
number of communities
detected for random graphs
of the same size (*blue*) and
degree sequence (*red*) as the
karate network

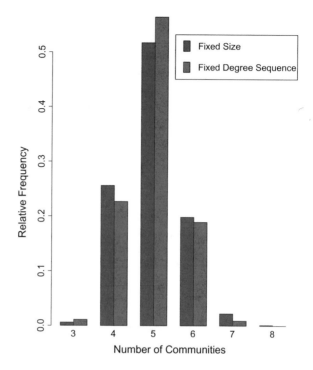

```
#5.29  1 > num.comm.grg <- numeric(ntrials)
       2 > for(i in (1:ntrials)){
       3 +   g.grg <- sample_degseq(degs, method="vl")
       4 +   c.grg <- cluster_fast_greedy(g.grg)
       5 +   num.comm.grg[i] <- length(c.grg)
       6 + }
```

The results may be summarized and compared using side by side bar plots.

```
#5.30  1 > rslts <- c(num.comm.rg,num.comm.grg)
       2 > indx <- c(rep(0, ntrials), rep(1, ntrials))
       3 > counts <- table(indx, rslts)/ntrials
       4 > barplot(counts, beside=TRUE, col=c("blue", "red"),
       5 +   xlab="Number of Communities",
       6 +   ylab="Relative Frequency",
       7 +   legend=c("Fixed Size", "Fixed Degree Sequence"))
```

See Fig. 5.6. Clearly the actual number of communities detected in the original karate
network (i.e., three) would be considered unusual from the perspective of random
graphs of both fixed size and fixed degree sequence. Accordingly, we may conclude
that there is likely an additional mechanism(s) at work in the actual karate club, one
that goes beyond simply the density and the distribution of social interactions in this
network.

5.5.2 Assessing Small World Properties

The notion of small-world networks has been particularly popular in the field of neuroscience, where numerous authors have presented evidence for arguments that small-world behavior can be found in various network-based representations of the brain. See [20] for a review. On the other hand, such claims are not completely uncontested, with recent work arguing that—in certain contexts—some or all of the observed behavior may be attributed to aspects of the sampling underlying the data used to construct such networks (e.g., [21, 22]).

A typical approach to assessing small-world behavior in this area is to compare the observed clustering coefficient and average (shortest) path length in an observed network to what might be observed in an appropriately calibrated classical random graph. Recalling our discussion of small world networks in Sect. 5.4.1, we should expect under such a comparison—if indeed an observed network exhibits small-world behavior—that the observed clustering coefficient exceed that of a random graph, while the average path length remain roughly the same.

The data set macaque in **igraph** contains a network of established functional connections between brain areas understood to be involved with the tactile function of the visual cortex in macaque monkeys [23].

```
#5.31  1 > library(igraphdata)
       2 > data(macaque)
       3 > summary(macaque)
       4 IGRAPH f7130f3 DN-- 45 463 --
       5 + attr: Citation (g/c), Author (g/c), shape (v/c),
       6   name (v/c)
```

It is a directed network of 45 vertices and 463 links.

In order to assess clustering in this network, we use an extension to directed graphs of the clustering coefficient defined in Sect. 4.3.2, due to [24]. This quantity is defined as the average, over all vertices v, of the vertex-specific clustering coefficients

$$\mathsf{cl}(v) = \frac{\left(\mathbf{A} + \mathbf{A}^T\right)^3_{vv}}{2\left[d_v^{tot}\left(d_v^{tot} - 1\right) - 2\left(\mathbf{A}^2\right)_{vv}\right]} \;, \tag{5.4}$$

where $\mathbf{A}$ is the adjacency matrix and d_v^{tot} is the total degree (i.e., in-degree plus out-degree) of vertex v. The calculation of this quantity is accomplished through the following function.

```
#5.32  1 > clust_coef_dir <- function(graph) {
       2 +   A <- as.matrix(as_adjacency_matrix(graph))
       3 +   S <- A + t(A)
       4 +   deg <- degree(graph, mode=c("total"))
       5 +   num <- diag(S %*% S %*% S)
       6 +   denom <- diag(A %*% A)
       7 +   denom <- 2 * (deg * (deg - 1) - 2 * denom)
       8 +   cl <- mean(num/denom)
```

```
 9 +       return(cl)
10 + }
```

Similarly, path lengths are defined in this network only for directed paths.

The steps required to simulate draws of (directed) classical random graphs, and to assess the clustering and average path length for each, are analogous to those employed in our example with the karate network just above.

```
#5.33  1 > ntrials <- 1000
       2 > nv <- vcount(macaque)
       3 > ne <- ecount(macaque)
       4 > cl.rg <- numeric(ntrials)
       5 > apl.rg <- numeric(ntrials)
       6 > for (i in (1:ntrials)) {
       7 +     g.rg <- sample_gnm(nv, ne, directed=TRUE)
       8 +     cl.rg[i] <- clust_coef_dir(g.rg)
       9 +     apl.rg[i] <- mean_distance(g.rg)
      10 + }
```

Summarizing the resulting distributions of clustering coefficient

```
#5.34  1 > summary(cl.rg)
       2    Min. 1st Qu.  Median    Mean 3rd Qu.    Max.
       3  0.2159  0.2302  0.2340  0.2340  0.2377  0.2548
```

and average path length

```
#5.35  1 > summary(apl.rg)
       2    Min. 1st Qu.  Median    Mean 3rd Qu.    Max.
       3   1.810   1.827   1.833   1.833   1.838   1.858
```

and comparing against these distributions the values for the macaque network

```
#5.36  1 > clust_coef_dir(macaque)
       2 [1] 0.5501073
       3 > mean_distance(macaque)
       4 [1] 2.148485
```

we find that our observed values fall far outside the range typical of these random graphs.

Therefore, we see that on the one hand there is substantially more clustering in our network than expected from a random network. On the other hand, however, the shortest paths between vertex pairs are, on average, also noticeably longer. Hence, the evidence for small-world behavior in this network is not clear, with the results suggesting that the network behaves more like a lattice than a classical random graph.

5.6   Additional Reading

There are a number of books devoted entirely to subsets of the topics in network graph modeling discussed in this chapter. For example, the volume by Bollobás [5] is the standard reference for classical random graph theory. See too the book by Janson, Luczak, and Rucinski [25]. For generalized random graph theory, see the books by Frieze and Karonski [26], van der Hofstad [27], or Chung and Lu [28]. Similarly, many books on network analysis include such material as well, to varying extents. See, for example, the book by Newman [29, Sect. 4].

References

1. P. Erdős and A. Rényi, "On random graphs," *Publ. Math. Debrecen*, vol. 6, no. 290, pp. 290–297, 1959.
2. P. Erdős and A. Rényi, "On the evolution of random graphs," *Publ. Math. Inst. Hung. Acad. Sci.*, vol. 5, pp. 17–61, 1960.
3. P. Erdős and A. Rényi, "On the strength of connectedness of a random graph," *Acta. Math. Acad. Sci. Hungar.*, vol. 12, pp. 261–267, 1961.
4. E. Gilbert, "Random graphs," *Annals of Mathematical Statistics*, vol. 30, no. 4, pp. 1141–1144, 1959.
5. E. Gilbert, *Random Graphs, Second Edition*. New York: Cambridge University Press, 2001.
6. T. Snijders, "Enumeration and simulation methods for $0 - 1$ matrices with given marginal totals," *Psychometrika*, vol. 57, pp. 397–417, 1991.
7. J. Roberts, "Simple methods for simulating sociomatrices with given marginal totals," *Social Networks*, vol. 22, no. 3, pp. 273–283, 2000.
8. J. McDonald, P. Smith, and J. Forster, "Markov chain Monte Carlo exact inference of social networks," *Social Networks*, vol. 29, pp. 127–136, 2007.
9. D. Watts and S. Strogatz, "Collective dynamics of 'small-world' networks," *Nature*, vol. 393, no. 6684, pp. 440–442, 1998.
10. A. Barabási and R. Albert, "Emergence of scaling in random networks," *Science*, vol. 286, no. 5439, pp. 509–512, 1999.
11. M. Mitzenmacher, "A brief history of generative models for power law and lognormal distributions," *Internet Mathematics*, vol. 1, no. 2, pp. 226–251, 2004.
12. S. Dorogovtsev, J. Mendes, and A. Samukhin, "Structure of growing networks with preferential linking," *Physical Review Letters*, vol. 85, no. 21, pp. 4633–4636, 2000.
13. P. Krapivsky, S. Redner, and F. Leyvraz, "Connectivity of growing random networks," *Physical Review Letters*, vol. 85, no. 21, pp. 4629–4632, 2000.
14. B. Bollobás, O. Riordan, J. Spencer, and G. Tusnady, "The degree sequence of a scale-free random graph process," *Random Structures and Algorithms*, vol. 18, no. 3, pp. 279–290, 2001.
15. B. Bollobás and O. Riordan, "The diameter of a scale-free random graph," *Combinatorica*, vol. 24, no. 1, pp. 5–34, 2004.
16. B. Bollobás and O. Riordan, "Mathematical results on scale-free random graphs," in *Handbook of Graphs and Networks: From the Genome to the Internet*, S. Bornholdt and H. Schuster, Eds. Weinheim: Wiley-VCH, 2002, pp. 1–34.
17. N. Kashtan, S. Itzkovitz, R. Milo, and U. Alon, "Efficient sampling algorithm for estimating subgraph concentrations and detecting network motifs," *Bioinformatics*, vol. 20, no. 11, pp. 1746–1758, 2004.
18. R. Milo, S. Shen-Orr, S. Itzkovitz, N. Kashtan, D. Chklovskii, and U. Alon, "Network motifs: simple building blocks of complex networks," *Science*, vol. 298, no. 5594, pp. 824–827, 2002.

19. E. Kolaczyk, *Statistical Analysis of Network Data: Methods and Models*. Springer Verlag, 2009.
20. D. S. Bassett and E. Bullmore, "Small-world brain networks," *The Neuroscientist*, vol. 12, no. 6, pp. 512–523, 2006.
21. S. Bialonski, M.-T. Horstmann, and K. Lehnertz, "From brain to earth and climate systems: Small-world interaction networks or not?" *Chaos: An Interdisciplinary Journal of Nonlinear Science*, vol. 20, no. 1, pp. 013 134–013 134, 2010.
22. F. Gerhard, G. Pipa, B. Lima, S. Neuenschwander, and W. Gerstner, "Extraction of network topology from multi-electrode recordings: is there a small-world effect?" *Frontiers in Computational Neuroscience*, vol. 5, 2011.
23. L. Négyessy, T. Nepusz, L. Kocsis, and F. Bazsó, "Prediction of the main cortical areas and connections involved in the tactile function of the visual cortex by network analysis," *European Journal of Neuroscience*, vol. 23, no. 7, pp. 1919–1930, 2006.
24. G. Fagiolo, "Clustering in complex directed networks," *Physical Review E*, vol. 76, no. 2, p. 026107, 2007.
25. S. Janson, T. Luczak, and A. Rucinski, *Random Graphs*. John Wiley & Sons, 2011, vol. 45.
26. A. Frieze and M. Karonski, *Introduction to Random Graphs*. Cambridge University Press, 2016.
27. R. Van Der Hofstad, *Random Graphs and Complex Networks*. Cambridge university press, 2016, vol. 1.
28. F. Chung and L. Lu, *Complex Graphs and Networks*. American Mathematical Society, 2006.
29. M. Newman, *Networks: an Introduction*. Oxford University Press, Inc., 2010.

Chapter 6
Statistical Models for Network Graphs

6.1 Introduction

The network models discussed in the previous chapter serve a variety of useful purposes. Yet for the purpose of statistical model building, they come up short. Indeed, as Robins and Morris [1] write, "A good [statistical network graph] model needs to be both estimable from data and a reasonable representation of that data, to be theoretically plausible about the type of effects that might have produced the network, and to be amenable to examining which competing effects might be the best explanation of the data." None of the models we have seen up until this point are really intended to meet such criteria.

In contrast, there are a number of other classes of network graph models which *are* designed explicitly for use as statistical models. In fact, the three main such classes of models developed to date closely parallel more familiar statistical models for non-network datasets. The class of exponential random graph models are analogous to standard regression models—particularly, generalized linear models. Similarly, stochastic block models draw their inspiration from mixture models, as they are, in their most basic form, essentially a mixture of classical random graph models. Finally, latent network models are a network-based variant of the common practice of using both observed and unobserved (i.e., latent) variables in modeling an outcome (i.e., in this case, the presence or absence of network edges).

It is important to note, however, that none of these models are simply direct implementations of their classical analogues. The adaptation of the latter to network-based data structures can have nontrivial implications on model specification and identifiability, model fitting, and the assessment of significance of terms in the model and model goodness of fit.

In this chapter we explore the basic structure and use of certain canonical examples of each of these three classes of statistical models for network graphs.

© Springer Nature Switzerland AG 2020

E. D. Kolaczyk and G. Csárdi, *Statistical Analysis of Network Data with R*, Use R!,
https://doi.org/10.1007/978-3-030-44129-6_6

6.2 Exponential Random Graph Models

Exponential random graph models (ERGMs)[1] are designed in direct analogy to the classical generalized linear models (GLMs). They are formulated in a manner that is intended to facilitate the adaptation and extension of well-established statistical principles and methods for the construction, fitting, and comparison of models. Nevertheless, the appropriate specification and fitting of ERGMs can be decidedly more subtle than with standard GLMs. Moreover, much of the standard inferential infrastructure available for GLMs, resting on asymptotic approximations to appropriate chi-square distributions, has yet be formally justified in the case of ERGMs. As a result, while this class of models arguably has substantial potential, in practice it must be used with some care.

6.2.1 General Formulation

Consider $G = (V, E)$ as a random graph. Let $Y_{ij} = Y_{ji}$ be a binary random variable indicating the presence or absence of an edge $e \in E$ between the two vertices i and j in V. The matrix $\mathbf{Y} = [Y_{ij}]$ is thus the (random) adjacency matrix for G. Denote by $\mathbf{y} = [y_{ij}]$ a particular realization of $\mathbf{Y}$. An exponential random graph model is a model specified in exponential family form[2] for the joint distribution of the elements in $\mathbf{Y}$. The basic specification for an ERGM is a model of the form

$$\mathbb{P}_\theta (\mathbf{Y} = \mathbf{y}) = \left(\frac{1}{\kappa}\right) \exp\left\{\sum_H \theta_H \, g_H(\mathbf{y})\right\} , \qquad (6.2)$$

where

(i) each H is a *configuration*, which is defined to be a set of possible edges among a subset of the vertices in G;

---

[1]These models have also been referred to as p^* models, particularly in the social network literature, where they are seen as one of the later examples of a series of model classes introduced in succession over a roughly 20-year period covering the late 1970s, 1980s, and early 1990s. See the review of Wasserman and Pattison [2], for example. Our use of the term 'exponential random graph models' reflects current practice, which emphasizes the connection of these models with traditional exponential family models in classical statistics.

[2]Recall that an arbitrary (discrete) random vector $\mathbf{Z}$ is said to belong to an *exponential family* if its probability mass function may be expressed in the form

$$\mathbb{P}_\theta (\mathbf{Z} = \mathbf{z}) = \exp\left\{\theta^T \mathbf{g}(\mathbf{z}) - \psi(\theta)\right\} , \qquad (6.1)$$

where $\theta \in \mathbb{R}^p$ is a $p \times 1$ vector of parameters, $\mathbf{g}(\cdot)$ is a p-dimensional function of $\mathbf{z}$, and $\psi(\theta)$ is a normalization term, ensuring that $\mathbb{P}_\theta(\cdot)$ sums to one over its range.

(ii) $g_H(\mathbf{y}) = \prod_{y_{ij} \in H} y_{ij}$, and is therefore either one if the configuration H occurs in $\mathbf{y}$, or zero, otherwise;

(iii) a non-zero value for θ_H means that the Y_{ij} are dependent for all pairs of vertices $\{i, j\}$ in H, conditional upon the rest of the graph; and

(iv) $\kappa = \kappa(\theta)$ is a normalization constant,

$$\kappa(\theta) = \sum_{\mathbf{y}} \exp\left\{ \sum_H \theta_H \, g_H(\mathbf{y}) \right\} . \tag{6.3}$$

Note that the summation in (6.2) is over all possible configurations H. Importantly, given a choice of functions g_H and their coefficients θ_H, this implies a certain (in)dependency structure among the elements in $\mathbf{Y}$, which is, of course, appealing, given the inherently relational nature of a network. Generally speaking, such structure typically can be described as specifying that the random variables $\{Y_{ij}\}_{(i,j)\in\mathscr{A}}$ are independent of $\{Y_{i'j'}\}_{(i',j')\in\mathscr{B}}$, conditional on the values of $\{Y_{i''j''}\}_{(i'',j'')\in\mathscr{C}}$, for some given index sets $\mathscr{A}, \mathscr{B}$, and $\mathscr{C}$. Conversely, we can begin with a collection of (in)dependence relations among subsets of elements in $\mathbf{Y}$ and try to derive the induced form of the (g_H, θ_H) pairs.[3]

The ERGM framework allows for a number of variations and extensions. For example, directed versions of ERGMs are also available. In addition, in defining ERGMs for either undirected or directed graphs, it is straightforward to include, if desired, information on vertices beyond their connectivity, such as actor attributes in a social network or known functionalities of proteins in a network of protein interactions. Given a realization $\mathbf{x}$ of a random vector $\mathbf{X}$ on the vertices in G, we simply specify an exponential form for the conditional distribution $\mathbb{P}_\theta(\mathbf{Y} = \mathbf{y}|\mathbf{X} = \mathbf{x})$ that involves additional statistics $g(\cdot)$ that are functions of both $\mathbf{y}$ and $\mathbf{x}$.

In this section, we will illustrate the construction, fitting, and assessment of ERGMs using the `lazega` data set on collaboration among lawyers, introduced in Chap. 1. Within R, easily the most comprehensive and sophisticated package for ERGMs is the **ergm** package, which is part of the **statnet** suite of packages.[4] Since **ergm** uses the **network** package to represent network objects, we convert the **igraph** object `lazega` to the format used in **statnet**, first separating the network into adjacency matrix and attributes

```
#6.1 1 > library(sand)
     2 > data(lazega)
     3 > A <- as_adjacency_matrix(lazega)
     4 > v.attrs <- as_data_frame(lazega, what="vertices")
```

and then creating the analogous network object for **ergm**

---

[3] However, it is important to realize that it is *not* the case that simply any collection of (in)dependence relations among the elements of $\mathbf{Y}$ yields a proper joint distribution on $\mathbf{Y}$. Rather, certain conditions must be satisfied, as formalized in the celebrated Hammersley-Clifford theorem (e.g., Besag [3]).

[4] The **statnet** suite is arguably the most sophisticated single collection of R packages for doing statistical modeling of network graphs, particularly from the perspective of social network analysis.

```
#6.2 1 > library(ergm)   # Will load package 'network' as well.
     2 > lazega.s <- network::as.network(as.matrix(A),
     3 +   directed=FALSE)
     4 > network::set.vertex.attribute(lazega.s, "Office",
     5 +    v.attrs$Office)
     6 > network::set.vertex.attribute(lazega.s, "Practice",
     7 +    v.attrs$Practice)
     8 > network::set.vertex.attribute(lazega.s, "Gender",
     9 +    v.attrs$Gender)
    10 > network::set.vertex.attribute(lazega.s, "Seniority",
    11 +    v.attrs$Seniority)
```

6.2.2 Specifying a Model

The general formulation just described leaves much flexibility in specifying an ERGM. We illustrate in the material that follows below, but refer the reader to, for example, the review article by Robins et al. [4] or the book by Lusher et al. [5] for a more comprehensive treatment.

We have already seen an example of what is arguably the simplest ERGM, in the form of the Bernoulli random graph model of Sect. 5.2. To see this, suppose we specify that, for a given pair of vertices, the presence or absence of an edge between that pair is independent of the status of possible edges between any other pairs of vertices.[5] Then $\theta_H = 0$ for all configurations H involving three or more vertices. As a result, the ERGM in (6.2) reduces to

$$\mathbb{P}_\theta (\mathbf{Y} = \mathbf{y}) = \left(\frac{1}{\kappa}\right) \exp \left\{ \sum_{i,j} \theta_{ij} y_{ij} \right\} . \tag{6.4}$$

Furthermore, if we assume that the coefficients θ_{ij} are equal to some common value θ (typically referred to as an assumption of *homogeneity* across the network), then (6.4) further simplifies to

$$\mathbb{P}_\theta (\mathbf{Y} = \mathbf{y}) = \left(\frac{1}{\kappa}\right) \exp \left\{ \theta L(\mathbf{y}) \right\} , \tag{6.5}$$

where $L(\mathbf{y}) = \sum_{i,j} y_{ij} = N_e$ is the number of edges in the graph. The result is equivalent to a Bernoulli random graph model, with $p = \exp(\theta)/[1 + \exp(\theta)]$.

To specify models in **ergm**, we use the function `formula` and standard R syntax. For example, model (6.5) may be specified for the network `lazega.s` as

```
#6.3 1 > my.ergm.bern <- formula(lazega.s ~ edges)
     2 > my.ergm.bern
     3 lazega.s ~ edges
```

in which case the statistic L takes the value

---

[5] That is, for each pair $\{i, j\}$, we assume that Y_{ij} is independent of $Y_{i',j'}$, for any $\{i', j'\} \neq \{i, j\}$.

```
#6.4 1 > summary(my.ergm.bern)
     2   edges
     3     115
```

The strength of ERGMs lies in our ability to specify decidedly more nuanced models than that above. Doing so properly and effectively, however, requires some thought and care.

To begin, note that the model in (6.5) can be thought of as specifying that the log-odds of observing a given network G (or, more specifically, its adjacency matrix $\mathbf{y}$) is simply proportional to the number of edges in the network—arguably the most basic of network statistics. Traditionally, it has been of interest to also incorporate analogous statistics of higher-order global network structure, such as counts of k-stars,[6] say $S_k(\mathbf{y})$, and of triangles, say $T(\mathbf{y})$. Frank and Strauss [6] show that models of the form

$$\mathbb{P}_\theta\left(\mathbf{Y} = \mathbf{y}\right) = \left(\frac{1}{\kappa}\right) \exp\left\{\sum_{k=1}^{N_v-1} \theta_k S_k(\mathbf{y}) + \theta_\tau T(\mathbf{y})\right\} \tag{6.6}$$

are equivalent to a certain limited form of dependence among the edges y_{ij}, in contrast to the independence specified by the Bernoulli model.[7]

In using such models, common practice has been to include star counts S_k no higher than $k = 2$, or at most $k = 3$, by setting $\theta_4 = \cdots = \theta_{N_v-1} = 0$. For example,

```
#6.5 1 > my.ergm <- formula(lazega.s ~ edges + kstar(2)
     2 +        + kstar(3) + triangle)
     3 > summary(my.ergm)
     4      edges    kstar2   kstar3 triangle
     5        115       926     2681      120
```

While simpler and, ideally, more interpretable, than the general formulation in (6.6), experience nevertheless has shown this practice to frequently produce models that fit quite poorly to real data. Investigation of this phenomena has found it to be intimately related to the issue of model degeneracy.[8] See Handcock [7]. Unfortunately, the alternative—including a sufficiently large number of higher order terms—is problematic as well, from the perspective of model fitting.

A solution to this dilemma, proposed by Snijders et al. [8], is to impose a parametric constraint of the form $\theta_k \propto (-1)^k \lambda^{2-k}$ upon the star parameters, for all $k \geq 2$, for some $\lambda \geq 1$. This tactic has the effect of combining all of the k-star statistics $S_k(\mathbf{y})$ in (6.6), for $k \geq 2$, into a single *alternating k-star statistic* of the form

---

[6]Note that $S_1(\mathbf{y}) = N_e$ is the number of edges.

[7]Formally, Frank and Strauss introduced the notion of *Markov dependence* for network graph models, which specifies that two possible edges are dependent whenever they share a vertex, conditional on all other possible edges. A random graph G arising under Markov dependence conditions is called a *Markov graph*.

[8] In this context the term is used to refer to a probability distribution that places a disproportionately large amount of its mass on a correspondingly small set of outcomes.

$$AKS_\lambda(\mathbf{y}) = \sum_{k=2}^{N_v-1} (-1)^k \frac{S_k(\mathbf{y})}{\lambda^{k-2}} \, , \tag{6.7}$$

and weighting that statistic by a single parameter θ_{AKS} that takes into account the star effects of all orders simultaneously. One may think of the alternating signs in (6.7) as allowing the counts of k-stars of successively greater order to balance each other, rather than simply ballooning (i.e., since more k-stars, for a given k, means more k'-stars, for $k' < k$).

Alternatively, and equivalently if the number of edges is included in the model, there is the *geometrically weighted degree count*, defined as

$$GWD_\gamma(\mathbf{y}) = \sum_{d=0}^{N_v-1} e^{-\gamma d} N_d(\mathbf{y}) \, , \tag{6.8}$$

where $N_d(\mathbf{y})$ is the number of vertices of degree d and $\gamma > 0$ is related to λ through the expression $\gamma = \log[\lambda/(\lambda - 1)]$. This approach in a sense attempts to model the degree distribution, with choice of γ influencing the extent to which higher-degree vertices are likely to occur in the graph G.

Snijders et al. [8] discuss a number of other similar statistics, including a generalization of triadic structures based on alternating sums of k-triangles, which takes the form[9]

$$AKT_\lambda(\mathbf{y}) = 3T_1 + \sum_{k=2}^{N_v-2} (-1)^{k+1} \frac{T_k(\mathbf{y})}{\lambda^{k-1}} \, . \tag{6.9}$$

Here T_k is the number of k-triangles, where a k-triangle is defined to be a set of k individual triangles sharing a common base. A discussion of the type of dependency properties induced among edges y_{ij} by such statistics can be found in Pattison and Robins [10].

These three statistics can be used in **ergm** by specifying terms altkstar, gwdegree, or gwesp, respectively, in the model. For example,

```
> my.ergm <- formula(lazega.s ~ edges
+    + gwesp(1, fixed=TRUE))
> summary(my.ergm)
        edges gwesp.fixed.1
     115.0000       213.1753
```

Note that all of the model specifications discussed so far involve statistics that are functions only of the network $\mathbf{y}$ (i.e., controlling for endogenous effects). Yet it is natural to expect that the chance of an edge joining two vertices depends not only on the status (i.e., presence or absence) of edges between other vertex pairs, but also on attributes of the vertices themselves (i.e., allowing for assessment of exogenous effects). For attributes that have been measured, we can incorporate them

---

[9]Hunter [9] offers an equivalent formulation of this definition, in terms of geometrically weighted counts of the neighbors common to adjacent vertices.

into the types of ERGMs we have seen, in the form of additional statistics in the exponential term in (6.2), with the normalization constant κ modified analogously, according to (6.3).

One natural form for such statistics is

$$g(\mathbf{y}, \mathbf{x}) = \sum_{1 \leq i < j \leq N_v} y_{ij}\, h(\mathbf{x}_i, \mathbf{x}_j)\ , \tag{6.10}$$

where h is a symmetric function of $\mathbf{x}_i$ and $\mathbf{x}_j$, and $\mathbf{x}_i$ (or $\mathbf{x}_j$) is the vector of observed attributes for the ith (or jth) vertex. Intuitively, if h is some measure of 'similarity' in attributes, then the statistic in (6.10) assesses the total similarity among network neighbors.

Two common choices of h produce analogues of 'main effects' and 'second-order effects' (or similarity or homophily effects) of certain attributes. Main effects, for a particular attribute x, are defined using a simple additive form:

$$h(x_i, x_j) = x_i + x_j\ . \tag{6.11}$$

On the other hand, second-order effects are defined using an indicator for equivalence of the respective attribute between two vertices, i.e.,

$$h(x_i, x_j) = I\{x_i = x_j\}\ . \tag{6.12}$$

Main effects and second-order effects may be incorporated into a model within **ergm** using the terms nodemain and nodematch, respectively.

To summarize, the various statistics introduced above have been chosen only to illustrate the many types of effects that may be captured in modeling network graphs using ERGMs. In modeling the network lazega.s throughout the rest of this section, we will draw on the analyses of Hunter and Handcock [11] and Snijders et al. [8]. In particular, we will specify a model of the form

$$\mathbb{P}_{\theta,\beta}(\mathbf{Y} = \mathbf{y}|\mathbf{X} = \mathbf{x}) = \left(\frac{1}{\kappa(\theta, \beta)}\right) \exp\left\{\theta_1\, S_1(\mathbf{y}) + \theta_2\, AKT_\lambda(\mathbf{y}) + \beta^T \mathbf{g}(\mathbf{y}, \mathbf{x})\right\}\ , \tag{6.13}$$

where $\mathbf{g}$ is a vector of five attribute statistics and β is the corresponding vector of parameters.

In R, our model is expressed as

```
#6.7  1  > lazega.ergm <- formula(lazega.s ~ edges
       2  +      + gwesp(log(3), fixed=TRUE)
       3  +      + nodemain("Seniority")
       4  +      + nodemain("Practice")
       5  +      + match("Practice")
       6  +      + match("Gender")
       7  +      + match("Office"))
```

This specification allows us to control for the density of the network and some effects of transitivity. In addition, it allows us to assess the effect on the formation of collaborative ties among lawyers that is had by seniority, the type of practice (i.e., corporate or litigation), and commonality of practice, gender, and office location.

6.2.3 Model Fitting

In standard settings, with independent and identically distributed realizations, exponential family models like that in (6.1) are generally fit using the method of maximum likelihood. In the context of the ERGMs in (6.2), the maximum likelihood estimators (MLEs) $\hat{\theta}_H$ of the parameters θ_H are well defined, assuming an appropriately-specified model, but their calculation is non-trivial.

Consider the general definition of an ERGM in (6.2). The MLE for the vector $\theta = (\theta_H)$ is defined as $\hat{\theta} = \arg\max_\theta \ell(\theta)$, where $\ell(\theta)$ is the log-likelihood, which has the particularly simple form common to exponential families,

$$\ell(\theta) = \theta^T \mathbf{g}(\mathbf{y}) - \psi(\theta) \ . \tag{6.14}$$

Here $\mathbf{g}$ denotes the vector of functions g_H and $\psi(\theta) = \log \kappa(\theta)$. Alternatively, taking derivatives on each side and using the fact that $\mathbb{E}_\theta[\mathbf{g}(\mathbf{Y})] = \partial \psi(\theta)/\partial \theta$, the MLE can also be expressed as the solution to the system of equations

$$\mathbb{E}_{\hat{\theta}}\left[\mathbf{g}(\mathbf{Y})\right] = \mathbf{g}(\mathbf{y}) \ . \tag{6.15}$$

Unfortunately, the function $\psi(\theta)$, occurring in both (6.14) and (6.15), cannot be evaluated explicitly in any but the most trivial of settings, as it involves the summation in (6.3) over $2^{\binom{N_v}{2}}$ possible choices of $\mathbf{y}$, for each candidate θ. Therefore, it is necessary to use numerical methods to compute approximate values for $\hat{\theta}$.

In **ergm**, models are fit using the function `ergm`, which implements a version of Markov chain Monte Carlo maximum likelihood estimation, deriving from the fundamental work of Geyer and Thompson [12]. See Hunter and Handcock [11], for example, for additional details and references. Our model in (6.13), for example, is fit as

```
#6.8 1 > set.seed(42)
     2 > lazega.ergm.fit <- ergm(lazega.ergm)
```

The analogy between ERGMs and GLMs may be drawn upon in summarizing and assessing the fit of the former.[10] For example, examination of an analysis of variance (ANOVA) table indicates that there is strong evidence that the variables used in the model `lazega.ergm` explain the variation in network connectivity to a highly nontrivial extent, with a change in deviance of 459 with only seven variables.

---

[10]We note that the **ergm** package provides not only summary statistics but also p-values. However, as mentioned earlier, the theoretical justification for the asymptotic chi-square and F-distributions used by **ergm** to compute these values has not been established to date. Therefore, our preference is to interpret these values informally, as additional summary statistics.

```
#6.9  1 > anova(lazega.ergm.fit)
      2 Analysis of Variance Table
      3
      4 Model 1: lazega.s ~ edges + gwesp(log(3), fixed = TRUE) +
      5     nodemain("Seniority") + nodemain("Practice") +
      6     match("Practice") + match("Gender") +
      7     match("Office")
      8           Df Deviance Resid. Df Resid. Dev Pr(>|Chisq|)
      9 NULL                       630      873.37
     10 Model 1:   7   413.74      623      459.63      < 2.2e-16 ***
     11 ---
     12 Signif. codes:  0  ***  0.001  **  0.01  *  0.05  .  0.1      1
```

Similarly, we can examine the relative contribution of the individual variables in our model.

```
#6.10  1 > summary(lazega.ergm.fit)
       2 ==========================
       3 Summary of model fit
       4 ==========================
       5
       6 Formula:   lazega.s ~ edges + gwesp(log(3), fixed = TRUE) +
       7     nodemain("Seniority") + nodemain("Practice") +
       8     match("Practice") + match("Gender") +
       9     match("Office")
      10
      11 Iterations:  2 out of 20
      12
      13 Monte Carlo MLE Results:
      14                             Estimate Std. Error MCMC %
      15 edges                       -7.00655    0.67114      0
      16 gwesp.fixed.1.09861228866811 0.59166    0.08554      0
      17 nodecov.Seniority            0.02456    0.00620      0
      18 nodecov.Practice             0.39455    0.10218      0
      19 nodematch.Practice           0.76966    0.19060      0
      20 nodematch.Gender             0.73767    0.24362      0
      21 nodematch.Office             1.16439    0.18753      0
      22
      23                               z value    Pr(>|z|)
      24 edges                         -10.440   < 1e-04 ***
      25 gwesp.fixed.1.09861228866811    6.917   < 1e-04 ***
      26 nodecov.Seniority               3.962   < 1e-04 ***
      27 nodecov.Practice                3.861   0.000113 ***
      28 nodematch.Practice              4.038   < 1e-04 ***
      29 nodematch.Gender                3.028   0.002463 **
      30 nodematch.Office                6.209   < 1e-04 ***
      31 ---
      32 Signif. codes:  0  ***  0.001  **  0.01  *  0.05  .  0.1      1
      33
      34     Null Deviance: 873.4  on 630  degrees of freedom
      35  Residual Deviance: 459.6  on 623  degrees of freedom
      36
      37 AIC: 473.6    BIC: 504.7    (Smaller is better.)
```

In order to interpret the coefficients, it is useful to think in terms of the probability of a given vertex pair having an edge, conditional on the edge status between all other pairs. Writing $\mathbf{Y}_{(-ij)}$ to be all of the elements of $\mathbf{Y}$ except Y_{ij}, the distribution of Y_{ij} conditional on $\mathbf{Y}_{(-ij)}$ is Bernoulli and satisfies the expression

$$\log\left[\frac{\mathbb{P}_{\theta}(Y_{ij} = 1 | \mathbf{Y}_{(-ij)} = \mathbf{y}_{(-ij)})}{\mathbb{P}_{\theta}(Y_{ij} = 0 | \mathbf{Y}_{(-ij)} = \mathbf{y}_{(-ij)})}\right] = \theta^T \Delta_{ij}(\mathbf{y}) , \qquad (6.16)$$

where $\Delta_{ij}(\mathbf{y})$ is the *change statistic*, denoting the difference between $\mathbf{g}(\mathbf{y})$ when $y_{ij} = 1$ and when $y_{ij} = 0$,

So the estimated coefficient of each attribute statistic in this analysis may be interpreted as a conditional log-odds ratio for cooperation between lawyers. For example, practicing corporate law, rather than litigation, increases the odds of cooperation by a factor of $\exp(0.3946) \approx 1.484$, or nearly 50%. Similarly, being of the same gender more than doubles the odds of cooperation, since $\exp(0.7377) \approx 2.091$. In all cases, such statements hold in the sense of 'all else being equal' (i.e., given no change among values of the other statistics). Note too that for all of the variables the coefficient differs from zero by at least one standard error, suggesting some nontrivial effect of these variables on the formation of network ties.

Similarly, in terms of network structure, the magnitude of the coefficient $\hat{\theta}_2 \approx 0.5917$ for the alternating k-triangle statistic and the comparatively small corresponding standard error indicate that there is also evidence for a nontrivial transitivity effect. Note that, given the inclusion of our second-order attribute statistics in the model, our quantification of this effect naturally controls for basic homophily on these attributes. So there is likely something other than similarity of gender, practice, and office at work here—possibly additional attributes we have not controlled for, or possibly social processes of team formation.

6.2.4 Goodness-of-Fit

In any sort of modeling problem, the best fit chosen from among a class of models need not necessarily be a *good* fit to the data if the model class itself does not contain a sufficiently rich set of models from which to choose. The concept of model *goodness-of-fit* is therefore important. But, while this concept is fairly well developed in standard modeling contexts, such as linear modeling, it arguably still has a good ways to go as far as network graph modeling is concerned.

For ERGMs, the current practice in assessing goodness-of-fit is to first simulate numerous random graphs from the fitted model and then compare various summaries of these graphs with those of the originally observed graph. If the characteristics of the observed network graph are too poor of a match to the typical values arising from realizations of the fitted random graph model, then this suggests systematic differences between the specified class of models and the data, and therefore a lack of goodness-of-fit.[11] In general, when assessing goodness-of-fit in network modeling, commonly used summaries include the distribution of any number of the various summaries of network structure encountered in Chap. 4, such as degree, centrality, and geodesic distance. With ERGMs, however, a natural choice of summary are the statistics g themselves that define the ERGM (i.e., the so-called sufficient statistics).

To assess the goodness-of-fit of our model in (6.13), as fit by `ergm`, the function `gof` in **ergm** runs the necessary Monte Carlo simulation and calculates comparisons with the original network graph in terms of the distributions of each of the seven statistics in the model.

```
#6.11 1 > gof.lazega.ergm <- gof(lazega.ergm.fit)
```

The results of these computations may then be plotted,

```
#6.12 1 > plot(gof.lazega.ergm)
```

as shown in Fig. 6.1. They indicate that—on the particular characteristics captured by these statistics—the fit of the model is quite good overall, with the observed statistics quite close to the median simulated values in most cases.

6.3 Network Block Models

We have seen that the structure of an ERGM closely parallels that of a standard regression model in statistics. The presence or absence of network edges (i.e., the Y_{ij}) is taken to be the response variable, while the role of the predictor variables is played by some combination of network summary statistics (i.e., endogenous variables) and functions of vertex and edge attributes (i.e., incorporating exogenous effects). In this section, we examine the class of network *block models*, which are instead analogous to classical mixture models.[12]

Recall that, in our analysis of the network of lawyer collaborations in the previous section, we used as predictors the sums of indicators that various attributes (e.g., practice or gender) were shared between vertex pairs. Importantly, while this choice may seem sensible from a practical perspective, it also reflects the potential impact on the formation of network ties of a key principle in social network theory—that of structural equivalence, i.e., the similarity of network positions and social roles. See [13, Chap. 9], for example. In general, we may think of vertices in a network as belonging to classes, and the propensity to establish ties between vertex pairs as depending on the class membership of the two vertices. With network block models these concepts are made precise.

---

[11]Goodness-of-fit has been found to be particularly important where ERGMs are concerned, due in large part to the issue of potential model degeneracy.

[12]A random variable X is said to follow a Q-class mixture distribution if its probability density function is of the form $f(x) = \sum_{q=1}^{Q} \alpha_q f_q(x)$, for class-specific densities f_q, where the mixing weights α_q are all non-negative and sum to one.

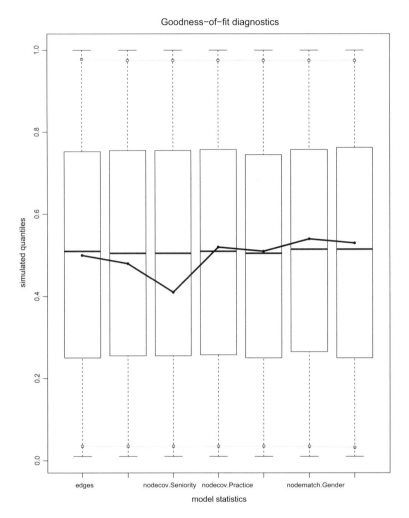

Fig. 6.1 Goodness-of-fit analysis, based on box-plots comparing the sufficient statistics of the original Lazega lawyer network and Monte Carlo realizations from the model in (6.13), with the parameters obtained by `ergm`

6.3.1 Model Specification

Suppose that each vertex $i \in V$ of a graph $G = (V, E)$ can belong to one of Q classes, say $\mathscr{C}_1, \ldots, \mathscr{C}_Q$. And furthermore, suppose that we know the class label $q = q(i)$ for each vertex i. A *block model* for G specifies that each element Y_{ij} of the adjacency matrix $\mathbf{Y}$ is, conditional on the class labels q and r of vertices i and j, respectively, an independent Bernoulli random variable, with probability π_{qr}. For an undirected graph, $\pi_{qr} = \pi_{rq}$.

The block model is hence a variant of the Bernoulli random graph model, where the probabilities of an edge are restricted to be one of only Q^2 possible values π_{qr}. Furthermore, in analogy to (6.5), this model can be represented in the form of an ERGM, i.e.,

$$\mathbb{P}_\theta\left(\mathbf{Y} = \mathbf{y}\right) = \left(\frac{1}{\kappa}\right) \exp\left\{\sum_{q,r} \theta_{qr} L_{qr}(\mathbf{y})\right\} , \tag{6.17}$$

where $L_{qr}(\mathbf{y})$ is the number of edges in the observed graph $\mathbf{y}$ connecting pairs of vertices of classes q and r.

Nevertheless, the assumption that the class membership of vertices is known or, moreover, that the 'true' classes $\mathscr{C}_1, \ldots, \mathscr{C}_Q$ have been correctly specified, is generally considered untenable in practice. More common, therefore, is the use of a *stochastic block model* (SBM) [14]. This model specifies only that there are Q classes, for some Q, but does not specify the nature of those classes nor the class membership of the individual vertices. Rather, it dictates simply that the class membership of each vertex i be determined independently, according to a common distribution on the set $\{1, \ldots, Q\}$.

Formally, let $Z_{iq} = 1$ if vertex i is of class q, and zero otherwise. Under a stochastic block model, the vectors $\mathbf{Z}_i = (Z_{i1}, \ldots, Z_{iQ})$ are determined independently, where $\mathbb{P}(Z_{iq} = 1) = \alpha_q$ and $\sum_{q=1}^{Q} \alpha_q = 1$. Then, conditional on the values $\{\mathbf{Z}_i\}$, the entries Y_{ij} are again modeled as independent Bernoulli random variables, with probabilities π_{qr}, as in the non-stochastic block model.

A stochastic block model is thus, effectively, a mixture of classical random graph models. As such, many of the properties of the random graphs G resulting from this model may be worked out in terms of the underlying model parameters. See [15], for example, who refer to this class of models as a 'mixture model for random graphs'.

Various extensions of the stochastic block model have in turn been proposed, although we will not pursue them here. For example, the class of mixed-membership stochastic block models allows vertices to be members of more than one class [16]. Similarly, the class of degree-corrected stochastic block models aims to produce mixtures of random graphs that have more heterogeneous degree distributions than the Poisson distribution corresponding to the classical random graph (e.g., [17, 18]).

6.3.2 Model Fitting

A non-stochastic block model can be fit in a straightforward fashion. The only parameters to be estimated are the edge probabilities π_{qr}, and the maximum likelihood estimates—which are natural here—are simply the corresponding empirical frequencies.

In the case of stochastic block models, both the (now conditional) edge probabilities π_{qr} and the class membership probabilities α_q must estimated. While this may not seem like much of a change over the ordinary block model, the task of model fitting becomes decidedly more complex in this setting. In order to see why, note that the log-likelihood for the joint distribution of the adjacency matrix $\mathbf{Y}$ and the class membership vectors $\{\mathbf{Z}_i\}$, i.e., the complete-data log-likelihood, is of the form

$$\ell(\mathbf{y}; \{\mathbf{z}_i\}) = \sum_i \sum_q z_{iq} \log \alpha_q + \frac{1}{2} \sum_{i \neq j} \sum_{q \neq r} z_{iq} z_{jr} \log b(y_{ij}; \pi_{qr}) , \qquad (6.18)$$

where $b(y; \pi) = \pi^y (1 - \pi)^{1-y}$. In principle, the likelihood of the observed data is obtained then by summing the complete-data likelihood over all possible values of $\{\mathbf{z}_i\}$. Unfortunately, to do so typically is intractable in problems of any real interest. As a result, computationally intensive methods must be used to produce estimates based on this likelihood.

The expectation-maximization (EM) algorithm [19] is a natural choice here. Effectively, given a current estimate of the π_{qr}, expected values of the $\mathbf{Z}_i$ are computed, conditional on $\mathbf{Y} = \mathbf{y}$. These values in turn are used to compute new estimates of the π_{qr}, using (conditional) maximum likelihood principles. The two steps are repeated in an alternating fashion, until convergence. But the first (i.e., expectation) step cannot be done in closed form, which greatly reduces the appeal of the algorithm.

Instead, a number of methods that approximate or alter the original maximum likelihood problem have been proposed in the literature. The R package **blockmodels** implements one such method (i.e., a so-called variational EM algorithm, which optimizes a lower bound on the likelihood of the observed data), for various types of stochastic block models.[13]

To illustrate, we use the network `fblog` of French political blogs introduced in Sect. 3.5. Recall that each blog is annotated as being associated with one of nine French political parties. Of course, these annotations do not necessarily correspond to an actual 'true' set of class groupings for these blogs, in the sense intended by the relatively simple form of the stochastic block model. Nevertheless, the context of the data (i.e., political blogs in the run-up to the French 2007 presidential election), as well as the various visualizations of this network in Sect. 3.5, suggest that it is likely the stochastic block model is not an unreasonable approximation to reality in this case.

Using the function `BM_bernoulli` in **blockmodels**, a fit to the observed network graph $\mathbf{y}$ is obtained through the following code.[14]

```
#6.13 1 > library(blockmodels)
      2 > set.seed(42)
      3 > A.fblog <- as.matrix(as_adjacency_matrix(fblog))
```

---

[13] In the first edition of this book, the R package **mixer** was featured for this purpose. At the time of this writing, **mixer** is no longer compliant with CRAN policy and has been archived as a result.

[14] Note that the call to `BM_bernoulli` creates an object of an S4 class type in R, rather than of the S3 type more typically found throughout this book. As a result, commands associated with manipulating the resulting **blockmodels** object `fblog.sbm` may be less familiar to some readers. For example, the command `fblog.sbm$estimate()` runs the default estimation procedure and then augments the initial object with the resulting output. A summary of the output may be obtained through the command `getClass(fblog.sbm)`.

```
4 > fblog.sbm <- BM_bernoulli("SBM_sym", A.fblog,
5 +                           verbosity=0, plotting='')
6 > fblog.sbm$estimate()
```

In fact, a number of stochastic block models are fit, over a range of values for Q, with the relevant output stored in lists indexed by Q. Comparison of these models, as a function of Q, is offered by BM_bernoulli through the so-called integration classification likelihood (ICL) criterion. This criterion is similar in spirit to various information criteria popular in standard regression modeling (e.g., Akaike's information (AIC), Bayesian information (BIC), etc.), but adapted specifically to clustering problems.

Based on the ICL criterion, we see that for the network of French blogs the optimal choice of Q is

```
#6.14 1 > ICLs <- fblog.sbm$ICL
      2 > Q <- which.max(ICLs)
      3 > Q
      4 [1] 10
```

classes. Focusing solely on the model with $Q = 10$ classes, we can extract the estimates of the posterior probability of class membership, i.e., the estimates of the expected values of the Z_i, conditional on $Y = y$.

```
#6.15 1 > Z <- fblog.sbm$memberships[[Q]]$Z
```

These estimates may, in turn, be used to determine labels for class assignments.

```
#6.16 1 > cl.labs <- apply(Z,1,which.max)
```

Thus stochastic block models may be used as a model-based method of graph partitioning, complementing the other methods introduced in Sect. 4.4. Note that in this case the evidence for class membership assignment appears to be uniformly strong (with maximum posterior probability exceeding 85%) across vertices.

```
#6.17 1 > nv <- vcount(fblog)
      2 > summary(Z[cbind(1:nv,cl.labs)])
      3    Min. 1st Qu.  Median    Mean 3rd Qu.    Max.
      4  0.8586  0.9953  0.9953  0.9938  0.9953  0.9953
```

It is also of interest to examine the parameter estimates associated with this model. For example, estimates of the class proportions α_q

```
#6.18 1 > cl.cnts <- as.vector(table(cl.labs))
      2 > alpha <- cl.cnts/nv
      3 > alpha
      4  [1] 0.18229167 0.14062500 0.05729167 0.10937500
      5  [5] 0.12500000 0.13020833 0.03125000 0.03645833
      6  [9] 0.06770833 0.11979167
```

reveal six larger classes and four smaller. Similarly, we see that, for example, within the third class, the blogs refer much more frequently to other blogs within the same class ($\pi_{33} > 0.9$) than to blogs in other classes ($\pi_{3r} < 0.05$ for $r \neq 3$).

```
#6.19 1 > Pi.mat <- fblog.sbm$model_parameters[[Q]]$pi
      2 > Pi.mat[3,]
      3  [1] 0.0030340287 0.0073657690 0.9102251927 0.0009221811
      4  [5] 0.0009170384 0.0364593875 0.0177621832 0.0024976022
      5  [9] 0.0431732528 0.0012495852
```

6.3.3 Goodness-of-Fit

In assessing the goodness-of-fit of a stochastic block model we could, of course, use
the same types of simulation-based methods we employed in the analysis of ERGMs
(i.e., illustrated in Fig. 6.1). The **igraph** function `sample_sbm` can be used for
this purpose. For example, a comparison of (i) the random graphs obtained through
simulation of our fitted stochastic block model `fblog.sbm` with (ii) the actual
French blog network, based on degree distribution, suggests that the two are quite
similar on average.

```
#6.20  1 > ntrials <- 1000
       2 > Pi.mat <- (t(Pi.mat)+Pi.mat)/2
       3 > deg.summ <- list(ntrials)
       4 > for(i in (1:ntrials)){
       5 +    blk.sz <- rmultinom(1,nv,alpha)
       6 +    g.sbm <- sample_sbm(nv,pref.matrix=Pi.mat,
       7 +                        block.sizes=blk.sz,
       8 +                        directed=FALSE)
       9 +    deg.summ[[i]] <- summary(degree(g.sbm))
      10 + }
      11 > Reduce('+',deg.summ)/ntrials
      12    Min. 1st Qu.  Median    Mean 3rd Qu.    Max.
      13   1.931   9.165  13.127  15.183  18.896  49.484
      14 > summary(degree(fblog))
      15    Min. 1st Qu.  Median    Mean 3rd Qu.    Max.
      16   2.00    8.00   13.00   14.91   18.00   56.00
```

However, the particular form of a stochastic block model lends itself as well to
certain other more model-specific devices. Two such visual summaries are displayed
in Fig. 6.2.

On the left is a plot of the ICL as a function of the number of classes Q.

```
#6.21  1 > plot(fblog.sbm$ICL,xlab="Q",ylab="ICL",type="b")
       2 > lines(c(Q,Q),c(min(ICLs),max(ICLs)),col="red",lty=2)
```

We see, for example, that while the fitted model has $Q = 10$ classes, the ICL criteria
seems to suggest there is perhaps some modest latitude in this choice. For these ten
classes, we can also produce a visualization of the adjacency matrix, with the rows
and columns regrouped by class, as shown on the right in the figure.

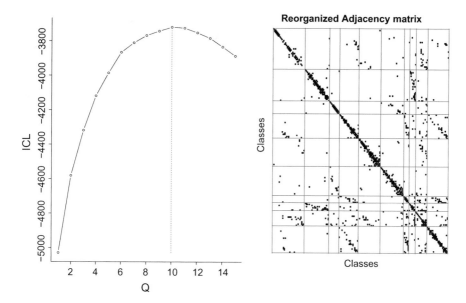

Fig. 6.2 Two plots summarizing the goodness-of-fit for the stochastic block model analysis of the French political blog network: integrated conditional likelihood (left) and reorganized adjacency matrix (right)

```
#6.22  1  > edges <- as_edgelist(fblog,names=FALSE)
       2  > neworder<-order(cl.labs)
       3  > m<-t(matrix(order(neworder)[as.numeric(edges)],2))
       4  > plot(1, 1, xlim = c(0, nv + 1), ylim = c(nv + 1, 0),
       5  +       type = "n", axes= FALSE, xlab="Classes",
       6  +       ylab="Classes",
       7  +       main="Reorganized Adjacency matrix")
       8  > rect(m[,2]-0.5,m[,1]-0.5,m[,2]+0.5,m[,1]+0.5,col=1)
       9  > rect(m[,1]-0.5,m[,2]-0.5,m[,1]+0.5,m[,2]+0.5,col=1)
      10  > cl.lim <- cl.cnts
      11  > cl.lim <- cumsum(cl.lim)[1:(length(cl.lim)-1)]+0.5
      12  > clip(0,nv+1,nv+1,0)
      13  > abline(v=c(0.5,cl.lim,nv+0.5),
      14  +         h=c(0.5,cl.lim,nv+0.5),col="red")
```

The clustering remarked upon earlier, into six larger classes and four smaller classes, is now evident to the eye. Furthermore, while it appears that the vertices in some of these classes are primarily connected with other vertices within their respective classes, among those other classes in which vertices show a propensity towards inter-class connections there seems to be, in some cases, a tendency towards connecting selectively with vertices of only certain other classes.

Finally, it is of interest to consider to what extent the graph partitioning induced by the vertex class assignments (i.e., into ten classes) matches the grouping of these blogs according to their political party status (i.e., according to nine parties). This comparison is summarized in Fig. 6.3.

```
#6.23  1 > g.cl <- graph_from_adjacency_matrix(Pi.mat,
       2 +                                      mode="undirected",
       3 +                                      weighted=TRUE)
       4 > # Set necessary parameters
       5 > vsize <- 100*sqrt(alpha)
       6 > ewidth <- 10*E(g.cl)$weight
       7 > PolP <- V(fblog)$PolParty
       8 > class.by.PolP <- as.matrix(table(cl.labs,PolP))
       9 > pie.vals <- lapply(1:Q, function(i)
      10 +                    as.vector(class.by.PolP[i,]))
      11 > my.cols <- topo.colors(length(unique(PolP)))
      12 > # Plot
      13 > plot(g.cl, edge.width=ewidth,
      14 +      vertex.shape="pie", vertex.pie=pie.vals,
      15 +      vertex.pie.color=list(my.cols),
      16 +      vertex.size=vsize, vertex.label.dist=0.1*vsize,
      17 +      vertex.label.degree=pi)
      18 > # Add a legend
      19 > my.names <- names(table(PolP))
      20 > my.names[2] <- "Comm. Anal."
      21 > my.names[5] <- "PR de G"
      22 > legend(x="topleft", my.names,
      23 +        fill=my.cols, bty="n")
      24
```

Here the circles, corresponding to the 10 vertex classes, and proportional in size to the number of blogs assigned to each class, are further broken down according to the relative proportion of political parties to which the blogs correspond, displayed in the form of pie charts. Connecting the circles are edges drawn with a thickness in proportion to the estimated probability that blogs in the two respective groups link to each other (i.e., in proportion to the estimated π_{qr}). Note that this plot may be contrasted with the coarse-level visualization of the original French blog network in Fig. 3.7.

A close examination of the pie charts yields, for example, that while the blogs in most of the 10 classes are quite homogeneous in their political party affiliations, two of the classes have a rather heterogeneous set of affiliations represented. In addition, certain of the political parties appear to be split between two or more classes. This latter observation might suggest that the model has chosen to use too many classes. Alternatively, it could instead indicate that there is actually splintering within these political parties.

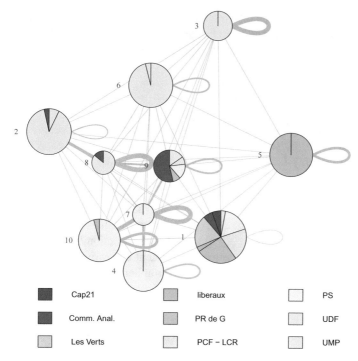

Fig. 6.3 Visual summary of the class connection probabilities π_{qr} for the network of French blogs and the concordance of those classes with party affiliations

6.4 Latent Network Models

From the perspective of statistical modeling, one key innovation underlying stochastic block models and their extensions is the incorporation of latent variables, in the form of vertex classes. That is, the use of variables that are unobserved but which play a role in determining the probability that vertex pairs are incident to each other. The principle of latent variables, common in many other areas of statistical modeling, has been adopted in a quite general sense with the class of *latent network models*. We draw on Hoff [20] in our development of these models below, and illustrate their usage with the R package **eigenmodel**, by the same author.

6.4.1 General Formulation

The incorporation of latent variables in network models for a random graph $G = (V, E)$ can be motivated by results of Hoover [21] and Aldous [22]. Specifically, in

the absence of any covariate information, the assumption of exchangeability[15] of the
vertices $v \in V$ is natural, and from this an argument can be made that each element
Y_{ij} of the adjacency matrix $\mathbf{Y}$ can be expressed in the form

$$Y_{ij} = h(\mu, u_i, u_j, \varepsilon_{ij}) , \qquad (6.19)$$

where μ is a constant, the u_i are independent and identically distributed latent variables, the ε_{ij} are independent and identically distributed pair-specific effects, and
the function h is symmetric in its second and third arguments. In other words, under
exchangeability, any random adjacency matrix $\mathbf{Y}$ can be written as a function of
latent variables.

Given the generality of the expression in (6.19), there are clearly many possible
latent network models we might formulate. If we specify that (i) the ε_{ij} are distributed
as standard normal random variables, (ii) the latent variables u_i, u_j enter into h only
through a symmetric function $\alpha(u_i, u_j)$, and (iii) the function h is simply an indicator
as to whether or not (i.e., one or zero) its argument is positive, and if in addition we
augment the parameter μ to include a linear combination of pair-specific covariates,
i.e., $\mathbf{x}_{ij}^T \beta$, then we arrive at a network version of a so-called probit model. Under this
model, the Y_{ij} are conditionally independent, with distributions

$$\mathbb{P}\left(Y_{ij} = 1 \mid \mathbf{X}_{ij} = \mathbf{x}_{ij}\right) = \Phi\left(\mu + \mathbf{x}_{ij}^T \beta + \alpha(u_i, u_j)\right) , \qquad (6.20)$$

where Φ is the cumulative distribution function of a standard normal random variable.[16]

If we denote the probabilities in (6.20) as p_{ij}, then the conditional model for $\mathbf{Y}$
as a whole takes the form

$$\mathbb{P}\left(\mathbf{Y} = \mathbf{y} \mid \mathbf{X}, u_1, \ldots, u_{N_v}\right) = \prod_{i<j} p_{ij}^{y_{ij}} (1 - p_{ij})^{1-y_{ij}} . \qquad (6.21)$$

That is, conditional on the covariates $\mathbf{X}$ and the latent variables $u_1, \ldots, u_{N_v}$, this
model for G has the form of a Bernoulli random graph model, with probabilities p_{ij}
specific to each vertex pair i, j. Note that complete specification of the full model
requires that a choice of distribution be made for the latent variables as well. We will
revisit this point later, in Sect. 6.4.3, after first exploring the issue of selecting the
form of the function $\alpha(\cdot, \cdot)$.

---

[15]A set of random variables is said to be *exchangeable* if their joint distribution is the same for any
ordering.

[16]In general, a probit model specifies, for a binary response Y, as a function of covariates $\mathbf{x}$, that
$\mathbb{P}(Y = 1|\mathbf{X} = \mathbf{x}) = \Phi(\mathbf{x}^T \beta)$, for some β.

6.4.2 Specifying the Latent Effects

The effect of the latent variables u on the probability of there being an edge between vertex pairs is largely dictated by the form of the function $\alpha(\cdot, \cdot)$. There have been a number of options explored in the literature to date. We remark briefly on three such options here.

A latent class model—analogous to the stochastic block models of Sect. 6.3—can be formulated by specifying that the u_i take values in the set $\{1, \ldots, Q\}$, and that $\alpha(u_i, u_j) = m_{u_i, u_j}$, for a symmetric matrix $\mathbf{M} = [m_{q,r}]$ of real-valued entries $m_{q,r}$. As remarked previously, the use of latent classes encodes into this model a notion of the principle of structural equivalence from social network theory.

Alternatively, the principle of homophily (i.e., the tendency of similar individuals to associate with each other) suggests an alternative choice, based on the concept of distance in a latent space. In this formulation, the latent variables u_i are simply vectors $(u_{i1}, \ldots, u_{iQ})^T$, of real numbers, interpreted as important but unknown characteristics of vertices that influence whether each establishes edges (e.g., social ties) with the others, and—importantly—vertices with more similar characteristics are expected to be more likely to establish an edge. Accordingly, the latent effects are specified as $\alpha(u_i, u_j) = -|u_i - u_j|$, for some distance metric $|\cdot|$, and the models are known as latent distance models.

Hoff [20] has suggested a third approach to specifying latent effects that combines the two approaches above, based on principles of eigen-analysis. Here the u_i are again Q-length random vectors, but the latent effects are given the form $\alpha(u_i, u_j) = u_i^T \Lambda u_j$, where Λ is a $Q \times Q$ diagonal matrix. Recall that the latent variables u are modeled as independent and identically distributed random vectors from the same distribution, and hence the correlation between each pair u_i and u_j is zero. While this is not the same as linear independence, in a linear algebraic sense, nevertheless it may be interpreted as saying that the u_i will be orthogonal 'in expectation'. Gathering the u_i into a matrix $\mathbf{U} = [u_1, \ldots, u_Q]$, the product $\mathbf{U} \Lambda \mathbf{U}^T$ therefore may be thought of as being in the spirit of an eigen-decomposition of the matrix of all pairwise latent effects $\alpha(u_i, u_j)$. Hoff refers to this model as an 'eigenmodel'.

The collection of eigenmodels can be shown to include the collection of latent class models in a formal sense, in that the set of matrices of latent effects that can be generated by the latter model is contained within that of the former model. In addition, there is a similar (albeit weaker) relationship between the collection of eigenmodels and the collection of latent distance models. As a result, the eigenmodel can be said to generalize both of these classes and, hence, its use allows for models that incorporate a blending of the principles of both structural equivalence and homophily. The manner in which the two principles are to be blended can be determined in a data-driven manner, through the process of model fitting.

6.4.3 Model Fitting

By construction, the latent network model has a hierarchical specification, so a Bayesian approach to inference is natural here. The package **eigenmodel** implements the eigenmodel formulation described above and will be the one with which we illustrate here.[17]

The function `eigenmodel_mcmc` in **eigenmodel** uses Monte Carlo Markov Chain (MCMC) techniques to simulate from the relevant posterior distributions, using largely conjugate priors to complete the model specification. Of potential interest are the parameter β (describing the effects of pair-specific covariates x_{ij}), the elements of the diagonal matrix Λ (summarizing the relative importance of each latent vector u_i), and the latent vectors themselves. Since the inferred latent vectors $\hat{u}_i$ are not orthogonal, it is useful in interpreting model output to use in their place the eigenvectors of the matrix $\hat{U} \, \hat{\Lambda} \, \hat{U}^T$.

The network `lazega` of collaborations among lawyers allows for demonstration of a number of the concepts we have discussed so far. Recall that this network involved 36 lawyers, at three different office locations, involved in two types of practice (i.e., corporate and litigation).

```
#6.24  1  > summary(lazega)
       2  IGRAPH NA UN-- 36 115 --
       3  + attr: name (v/c), Seniority (v/n), Status (v/n),
       4       Gender (v/n), Office (v/n), Years (v/n), Age (v/n),
       5       Practice (v/n), School (v/n)
```

We might hypothesize that collaboration in this setting is driven, at least in part, by similarity of practice, a form of homophily. On the other hand, we could similarly hypothesize that collaboration is instead driven by shared office location, which could be interpreted as a proxy for distance. Because the eigenmodel formulation of latent network models is able to capture aspects of both distance and homophily, it is interesting to compare the fitted models that we obtain for three different eigenmodels, specifying (i) no pair-specific covariates, (ii) a covariate for common practice, and (iii) a covariate for shared office location, respectively.

To fit the model with no pair-specific covariates and a latent eigen-space of $Q = 2$ dimensions is accomplished as follows.[18]

```
#6.25  1  > library(eigenmodel)
       2  > set.seed(42)
       3  > A <- as_adjacency_matrix(lazega, sparse=FALSE)
       4  > lazega.leig.fit1 <- eigenmodel_mcmc(A, R=2, S=11000,
       5  +    burn=10000)
```

---

[17]The package **latentnet**, in the **statnet** suite of tools, implements other variants of latent network models, such as latent distance models.

[18]The arguments S and burn chosen in our example ask that a 'burn-in' of $10,000$ iterations be used to initiate our MCMC sampler, after which the following $1,000$ iterations are used to perform posterior inference.

In order to include the effects of common practice, we create an array with that information

```
#6.26 1 > same.prac.op <- v.attr.lazega$Practice \%o\%
      2 +     v.attr.lazega$Practice
      3 > same.prac <- matrix(as.numeric(same.prac.op
      4 +   \%in\% c(1, 4, 9)), 36, 36)
      5 > same.prac <- array(same.prac,dim=c(36, 36, 1))
```

and fit the model with this additional argument

```
#6.27 1 > lazega.leig.fit2 <- eigenmodel_mcmc(A, same.prac, R=2,
      2 +     S=11000,burn=10000)
```

Finally, we do similarly for the model that includes information on shared office locations.

```
#6.28 1 > same.off.op <- v.attr.lazega$Office \%o\%
      2 +     v.attr.lazega$Office
      3 > same.off <- matrix(as.numeric(same.off.op \%in\%
      4 +   c(1, 4, 9)), 36, 36)
      5 > same.off <- array(same.off,dim=c(36, 36, 1))
      6 > lazega.leig.fit3 <- eigenmodel_mcmc(A, same.off,
      7 +     R=2, S=11000, burn=10000)
```

In order to compare the representation of the network `lazega` in each of the underlying two-dimensional latent spaces inferred for these models, we extract the eigen-
vectors for each fitted model

```
#6.29 1 > lat.sp.1 <-
      2 +     eigen(lazega.leig.fit1$ULU_postmean)$vec[, 1:2]
      3 > lat.sp.2 <-
      4 +     eigen(lazega.leig.fit2$ULU_postmean)$vec[, 1:2]
      5 > lat.sp.3 <-
      6 +     eigen(lazega.leig.fit3$ULU_postmean)$vec[, 1:2]
```

and plot the network in **igraph** using these coordinates as the layout.[19] For example,

```
#6.30 1 > colbar <- c("red", "dodgerblue", "goldenrod")
      2 > v.colors <- colbar[V(lazega)$Office]
      3 > v.shapes <- c("circle", "square")[V(lazega)$Practice]
      4 > v.size <- 3.5*sqrt(V(lazega)$Years)
      5 > v.label <- V(lazega)$Seniority
      6 > plot(lazega, layout=lat.sp.1, vertex.color=v.colors,
      7 +       vertex.shape=v.shapes, vertex.size=v.size,
      8 +       vertex.label=(1:36))
```

generates the visualization corresponding to the fit without any pair-specific covariates, and those for the other two models are obtained similarly (Fig. 6.4).

Examination of these three visualizations indicates that while the first two are
somewhat similar, the third is distinct. In particular, while the lawyers in the first two

---

[19]Conventions of vertex color, shape, and label are the same as in Fig. 1.1 in Chap. 1, and are
specified in R in the same manner as seen in Chap. 3.

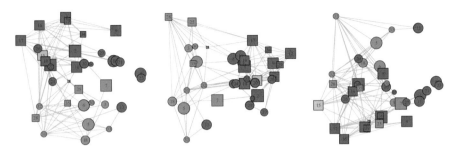

Fig. 6.4 Visualizations of the network of Lazega's lawyers, with layouts determined according to the inferred latent eigenvectors in models with no pair-specific covariates (*left*), a covariate for common practice (*center*), and a covariate for shared office location (*right*)

visualizations appear to be clustered into two main groups distinguished largely by common office location (i.e., color), in the third there appears to be only one main cluster. These observations suggest that common practice explains comparatively much less coarse-scale network structure than shared office location. And, indeed, when shared office location is taken into account, there appears to be decidedly less structure left to be captured by the latent variables.

6.4.4   Goodness-of-Fit

Here again, in assessing the goodness-of-fit of a latent network model we could, of course, use the same types of simulation-based methods we employed in the analysis of ERGMs (i.e., illustrated in Fig. 6.1). Alternatively, a more global sense of goodness-of-fit can be obtained by using principles of cross-validation. Specifically, a common practice in network modeling is to assess the accuracy with which, in fitting a model to a certain subset of the network, the remaining part of the network may be predicted. This notion usually is implemented through K-fold cross-validation, wherein the observed values y_{ij} are partitioned into K subsets (e.g., $K = 5$ is a standard choice), and the values in those subsets are predicted after training the same model on each of the complements of those subsets.

For example, consider the model fit above to the data lazega with no pair-specific covariates. After initiating a permutation of the $36 \times 35/2 = 630$ unique off-diagonal elements of the symmetric adjacency matrix, and initializing vector-based representations of the corresponding lower triangular portion of this matrix

```
#6.31 1 > perm.index <- sample(1:630)
      2 > nfolds <- 5
      3 > nmiss <- 630/nfolds
      4 > Avec <- A[lower.tri(A)]
      5 > Avec.pred1 <- numeric(length(Avec))
```

the process of cross-validation is implemented in the following lines.

```
#6.32  1 > for(i in seq(1,nfolds)){
       2 >   # Index of missing values.
       3 >   miss.index <- seq(((i-1) * nmiss + 1),
       4 +      (i*nmiss), 1)
       5 >   A.miss.index <- perm.index[miss.index]
       6 >
       7 >   # Fill a new Atemp appropriately with NA's.
       8 >   Avec.temp <- Avec
       9 >   Avec.temp[A.miss.index] <-
      10 +      rep("NA", length(A.miss.index))
      11 >   Avec.temp <- as.numeric(Avec.temp)
      12 >   Atemp <- matrix(0, 36, 36)
      13 >   Atemp[lower.tri(Atemp)] <- Avec.temp
      14 >   Atemp <- Atemp + t(Atemp)
      15 >
      16 >   # Now fit model and predict.
      17 >   Y <- Atemp
      18 >
      19 >   model1.fit <- eigenmodel_mcmc(Y, R=2,
      20 +      S=11000, burn=10000)
      21 >   model1.pred <- model1.fit$Y_postmean
      22 >   model1.pred.vec <-
      23 +      model1.pred[lower.tri(model1.pred)]
      24 >   Avec.pred1[A.miss.index] <-
      25 +      model1.pred.vec[A.miss.index]
      26 > }
```

Similarly, we can do the same for the models fit above with pair-specific covariates for common practice and shared office location, respectively, yielding, say, `Avec.pred2` and `Avec.pred3`. The results of the predictions generated under each of these three models can be assessed by examination of the corresponding receiver operating characteristic (ROC) curves.[20]

For example, using the package **ROCR**, an ROC curve for the predictions based on our first model are generated as follows.

```
#6.33  1 > library(ROCR)
       2 > pred1 <- prediction(Avec.pred1, Avec)
       3 > perf1 <- performance(pred1, "tpr", "fpr")
       4 > plot(perf1, col="blue", lwd=3)
```

The ROC curves for each of our three latent network models for the Lazega lawyer network are shown in Fig. 6.5.

---

[20]An ROC curve is used commonly in classification problems. The term refers to a curve obtained by plotting the true positive rate of a classifier against the true negative rate, as a threshold (or similar parameter) is varied across its natural range, where the threshold is applied to the predicted values to discriminate between two classes of interest. Here, since the predictions are posterior probabilities, the threshold is varied from 0 to 1, with vertex pairs for which the posterior probability of an edge is above threshold being predicted to have an edge.

Fig. 6.5 ROC curves comparing the goodness-of-fit to the Lazega network of lawyer collaboration for three different eigenmodels, specifying (i) no pair-specific covariates (*blue*), (ii) a covariate for common practice (*red*), and (iii) a covariate for shared office location (*yellow*), respectively

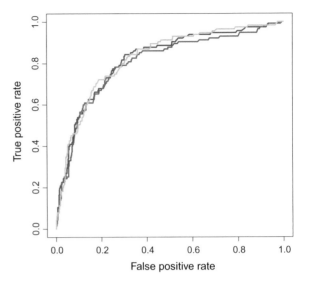

We see that from the perspective of predicting edge status, all three models appear to be comparable in their performance and to perform reasonably well, with an area under the curve (AUC) of roughly 80%.

```
#6.34  1 > perf1.auc <- performance(pred1, "auc")
       2 > slot(perf1.auc, "y.values")
       3 [[1]]
       4 [1] 0.8191811
```

6.5   Additional Reading

Of the three model classes discussed in this chapter, ERGMs have by far the longest and most extensive development, which has been summarized in the review article by Robins et al. [4] and detailed in the book by Lusher et al. [5]. For additional details on latent network models—in particular, stochastic block models—see the recent summary by Kolaczyk [23, Chap. 2].

References

1. G. Robins and M. Morris, "Advances in exponential random graph (p^*) models," *Social Networks*, vol. 29, no. 2, pp. 169–172, 2007.
2. S. Wasserman and P. Pattison, "Logit models and logistic regressions for social networks: I. An introduction to Markov graphs and p^*," *Psychometrika*, vol. 61, no. 3, pp. 401–425, 1996.
3. J. Besag, "Spatial interaction and the statistical analysis of lattice systems," *Journal of the Royal Statistical Society, Series B*, vol. 36, no. 2, pp. 192–236, 1974.

4. G. Robins, P. Pattison, Y. Kalish, and D. Lusher, "An introduction to exponential random graph (p^*) models for social networks," *Social networks*, vol. 29, no. 2, pp. 173–191, 2007.
5. D. Lusher, J. Koskinen, and G. Robins, *Exponential Random Graph Models for Social Networks: Theory, Methods, and Applications*. Cambridge University Press, 2012.
6. O. Frank and D. Strauss, "Markov graphs," *Journal of the American Statistical Association*, vol. 81, no. 395, pp. 832–842, 1986.
7. M. Handcock, "Assessing degeneracy in statistical models of social networks," Center for Statistics and the Social Sciences, University of Washington, Tech. Rep. No. 39, 2003.
8. T. Snijders, P. Pattison, G. Robins, and M. Handcock, "New specifications for exponential random graph models," *Sociological Methodology*, vol. 36, no. 1, pp. 99–153, 2006.
9. D. Hunter, "Curved exponential family models for social networks," *Social Networks*, vol. 29, no. 2, pp. 216–230, 2007.
10. P. Pattison and G. Robins, "Neighborhood-based models for social networks," *Sociological Methodology*, vol. 32, no. 1, pp. 301–337, 2002.
11. D. Hunter and M. Handcock, "Inference in curved exponential family models for networks," *Journal of Computational and Graphical Statistics*, vol. 15, no. 3, pp. 565–583, 2006.
12. C. Geyer and E. Thompson, "Constrained Monte Carlo maximum likelihood for dependent data," *Journal of the Royal Statistical Society, Series B*, vol. 54, no. 3, pp. 657–699, 1992.
13. S. Wasserman and K. Faust, *Social Network Analysis: Methods and Applications*. New York: Cambridge University Press, 1994.
14. K. Nowicki and T. Snijders, "Estimation and prediction for stochastic blockstructures," *Journal of the American Statistical Association*, vol. 96, no. 455, pp. 1077–1087, 2001.
15. J.-J. Daudin, F. Picard, and S. Robin, "A mixture model for random graphs," *Statistics and Computing*, vol. 18, no. 2, pp. 173–183, 2008.
16. E. M. Airoldi, D. M. Blei, S. E. Fienberg, and E. P. Xing, "Mixed membership stochastic blockmodels," *The Journal of Machine Learning Research*, vol. 9, pp. 1981–2014, 2008.
17. A. Coja-Oghlan and A. Lanka, "Finding planted partitions in random graphs with general degree distributions," *SIAM Journal on Discrete Mathematics*, vol. 23, no. 4, pp. 1682–1714, 2009.
18. B. Karrer and M. E. Newman, "Stochastic blockmodels and community structure in networks," *Physical Review E*, vol. 83, no. 1, p. 016107, 2011.
19. G. McLachlan and T. Krishnan, *The EM algorithm and extensions*. John Wiley & Sons, 2007, vol. 382.
20. G. McLachlan and T. Krishnan, "Modeling homophily and stochastic equivalence in symmetric relational data," *Advances in Neural Information Processing Systems, NIPS*, 2008.
21. D. N. Hoover, "Row-column exchangeability and a generalized model for probability," *Exchangeability in Probability and Statistics, North-Holland, Amsterdam*, pp. 81–291, 1982.
22. D. Aldous, "Exchangeability and related topics," *École d'Été de Probabilités de Saint-Flour XIII-1983*, pp. 1–198, 1985.
23. E. D. Kolaczyk, *Topics at the Frontier of Statistics and Network Analysis:(re) Visiting the Foundations*. Cambridge University Press, 2017.

Chapter 7
Network Topology Inference

7.1 Introduction

Network graphs are constructed in all sorts of ways and to varying levels of completeness. In some settings, there is little if any uncertainty in assessing whether or not an edge exists between two vertices and we can exhaustively assess incidence between vertex pairs. For example, in examining one's own network of Facebook friends, the presence or absence of an edge can be assessed through direct inspection. In other settings, however, constructing a network graph is not so straightforward. We may have information only on the status of some of the potential edges in the network, but not all. Alternatively, we may not even have the ability to directly assess whether or not an edge is present. Rather, it may be that we can only measure vertex or edge attributes that are to some extent predictive of edge status. In such cases, it can be natural to consider the task of constructing a network graph representation from the available data as one of statistical inference.

To be more precise, suppose that, broadly speaking, we have a set of measurements from a system of interest, such as vertex attributes $\mathbf{x} = (x_1, \ldots, x_{N_v})^T$ or binary indicators $\mathbf{y} = [y_{ij}]$ of certain edges but not others, or both $\mathbf{x}$ and $\mathbf{y}$, and we have a collection $\mathcal{G}$ of potential network graphs G. We might then take as our goal to select an appropriate member of $\mathcal{G}$ that best captures the underlying state of the system, based upon the information in the data as well as any other prior information, using techniques of statistical modeling and inference. That is, we might pose the problem as one of *network topology inference*.

In this chapter, we will focus on three particular classes of problems in network topology inference. Each class of problems is somewhat 'canonical' in nature, being easily posed in more than just a single specific network context. More specifically, in Sect. 7.2 we consider the problem of inferring whether or not a pair of vertices does or does not have an edge between them (i.e., inference of 'edge' or 'non-edge' status), using measurements that include a subset of vertex pairs whose edge/non-edge status is already observed. This problem is commonly referred to as link prediction. Next, in Sect. 7.3, we discuss the inference of association networks. Here the relation defining

© Springer Nature Switzerland AG 2020

E. D. Kolaczyk and G. Csárdi, *Statistical Analysis of Network Data with R*, Use R!,
https://doi.org/10.1007/978-3-030-44129-6_7

edges is taken by definition to be a nontrivial level of association (e.g., correlation) between certain characteristics of the vertices, but is itself unobserved and must be inferred from measurements reflecting these characteristics. Finally, we examine problems of tomographic network inference briefly in Sect. 7.4. These problems are distinguished by the fact that measurements are available only at vertices that are somehow at the 'perimeter' of the network, and it is necessary to infer the presence or absence of both edges and vertices in the 'interior.'

The ordering of these three classes of problems has a general progression. The first assumes knowledge of all of the vertices and the status of some of the edges/non-edges of the network graph, and seeks to infer the rest of the edges/non-edges. The second starts with no knowledge of edge status anywhere in the network graph, but assumes relevant measurements at all of the vertices and seeks to infer edge status throughout the network using these measurements. The third involves measurements at only a particular subset of vertices, which nevertheless indirectly provide some information useful for inferring the unknown topology of the rest of the network graph. A visual characterization of these three types of problems is shown in Fig. 7.1.

7.2 Link Prediction

Let $\mathbf{Y}$ be the random $N_v \times N_v$ binary adjacency matrix for a network graph $G = (V, E)$. Suppose that we observe only some of the entries of $\mathbf{Y}$, while others are missing. Denote the observed and missing elements of $\mathbf{Y}$ as $\mathbf{Y}^{obs}$ and $\mathbf{Y}^{miss}$, respectively. The problem of *link prediction* is to predict the entries in $\mathbf{Y}^{miss}$, given the values $\mathbf{Y}^{obs} = \mathbf{y}^{obs}$ and possibly various vertex attribute variables $\mathbf{X} = \mathbf{x} = (x_1, \ldots, x_{N_v})^T$. In other words, we wish to predict whether 'potential edges' between pairs of vertices in a network graph are present or absent using information provided by a subset of observed edges/non-edges and, if available, vertex attributes.

There are a number of variants of the link prediction problem that have been formulated, arising in settings ranging from information networks (e.g., [1–3]), to social networks (e.g., [4]), to biomolecular networks (e.g., [5, 6]). Besides differing in context, these variants of the problem can also differ, importantly, in why and how the values in $\mathbf{Y}^{miss}$ are missing. Sometimes there is simply a temporal component to the problem and, for example, edges may be 'missing' only in the sense that they are absent up to a certain point in time and then become present, as in Liben-Nowell and Kleinberg [1] and their study of the growth of the World Wide Web network graph. In many cases, however, all edges and non-edges effectively coincide in time, but the status of potential edges is missing due to issues of sampling. In this latter case the underlying mechanism of missingness can be important.

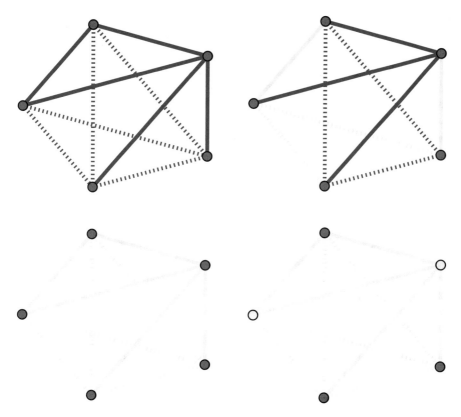

Fig. 7.1 Visual characterization of three types of network topology inference problems, for a toy network graph G. Edges shown in *solid*; non-edges, *dotted*. Observed vertices and edges shown in *dark* (i.e., *red* and *blue*, respectively); un-observed vertices and edges, in *light* (i.e., *pink* and *light blue*). *Top left*: True underlying graph G. *Top right*: Link prediction. *Bottom left*: Association network inference. *Bottom right*: Tomographic network inference

A common assumption (e.g., [3, 4]), and one we shall make here as well, is that the missing information on edge presence/absence is *missing at random*. This assumption means, essentially, that the probability that an edge variable Y_{ij} is observed depends only on the values of those other edge variables observed and not, for example, on its own value.[1]

Given an appropriate model for $\mathbf{X}$ and $(\mathbf{Y}^{obs}, \mathbf{Y}^{miss})$, such as those discussed in Chap. 6, we might aim to jointly predict the elements of $\mathbf{Y}^{miss}$ based on the induced model for

$$\mathbb{P}\left(\mathbf{Y}^{miss} \mid \mathbf{Y}^{obs} = \mathbf{y}^{obs}, \mathbf{X} = \mathbf{x}\right) . \tag{7.1}$$

---

[1] See Little and Rubin [7], for example, for a more formal definition and a general introduction to such missing data concepts.

But serious pursuit of this strategy entails, in part, successfully meeting the various challenges of modeling network graphs already described in Chap. 6, and in addition modeling the missingness in an appropriate fashion.

Perhaps not surprisingly, therefore, to date most model-based efforts in this area have instead focused upon the comparatively more manageable task of predicting the variables Y_{ij}^{miss} individually. Not only does this approach simplify matters considerably in many ways, but it also appears to in fact be a strong competitor to methods that predict the variables Y_{ij}^{miss} jointly. See Taskar et al. [3], for example.

In fact, we have already seen this approach demonstrated in the use of cross-validation as a framework for assessing model goodness-of-fit, in Sect. 6.4. There the values that we leave out for each of the K folds are indeed missing-at-random, by design. And the predictions for the status of each potential edge between vertex pairs i and j were based on the posterior expectation of Y_{ij} under our model.

Alternatively, scoring methods—methods based on the use of score functions—although somewhat less formal than model-based approaches, are nevertheless popular and can be quite effective. With scoring methods, for each pair of vertices i and j whose edge status is unknown, a score $s(i, j)$ is computed. A set of predicted edges may then be returned either by applying a threshold s^* to these scores, for some fixed s^*, or by ordering them and keeping those pairs with the top n^* values, for some fixed n^*.

There are many types of scores that have been proposed in the literature. They generally are designed to assess certain structural characteristics of a network graph $G^{obs} = (V^{obs}, E^{obs})$ associated with $\mathbf{Y}^{obs} = \mathbf{y}^{obs}$. A simple score, inspired by the 'small-world' principle, is negative the shortest-path distance between i and j,

$$s(i, j) = -\,\mathsf{dist}_{G^{obs}}(i, j)\ . \tag{7.2}$$

The negative sign in (7.2) is present so as to have larger score values indicate vertex pairs more likely to share an edge.

There are also a number of scores based on comparison of the observed neighborhoods $\mathcal{N}_i^{obs}$ and $\mathcal{N}_j^{obs}$ of i and j in G^{obs}, the simplest being the number of common neighbors

$$s(i, j) = |\mathcal{N}_i^{obs} \cap \mathcal{N}_j^{obs}|\ . \tag{7.3}$$

To illustrate the potential of simple scoring methods like this, recall the network `fblog` of French political blogs. The number of nearest common neighbors for each pair of vertices in this network, excluding—if incident to each other—the two vertices themselves, may be computed in the following manner.

```
#7.1  1  > library(sand)
       2  > nv <- vcount(fblog)
       3  > ncn <- numeric()
       4  > A <- as_adjacency_matrix(fblog)
       5  >
       6  > for(i in (1:(nv-1))){
       7  +    ni <- ego(fblog, 1, i)
```

```
 8 +    nj <- ego(fblog, 1, (i+1):nv)
 9 +    nbhd.ij <- mapply(intersect, ni, nj, SIMPLIFY=FALSE)
10 +    temp <- unlist(lapply(nbhd.ij, length)) -
11 +       2*A[i, (i+1):nv]
12 +    ncn <- c(ncn, temp)
13 + }
```

In Fig. 7.2 we compare the scores $s(i, j)$ for those vertex pairs that are not incident to each other (i.e., 'no edge'), and those that are (i.e., 'edge'), using so-called violin plots.[2]

```
#7.2 1 > library(vioplot)
     2 > Avec <- A[lower.tri(A)]
     3 > vioplot(ncn[Avec==0], ncn[Avec==1],
     4 +    names=c("No Edge", "Edge"),
     5 +    col="magenta")
     6 > title(ylab="Number of Common Neighbors")
```

It is evident from this comparison that there is a decided tendency towards larger scores when there is in fact an edge present. Viewing the calculation we have done here as a 'leave-one-out' cross-validation, and calculating the area under the curve (AUC) of the corresponding ROC curve,

```
#7.3 1 > library(ROCR)
     2 > pred <- prediction(ncn, Avec)
     3 > perf <- performance(pred, "auc")
     4 > slot(perf, "y.values")
     5 [[1]]
     6 [1] 0.9275179
```

we obtain further confirmation of the power of this score statistic to discriminate between edges and non-edges in this network.

Other choices of similar score statistics include a standardized version of this value, called the *Jaccard coefficient*,

$$s(i, j) = \frac{|\mathcal{N}_i^{obs} \cap \mathcal{N}_j^{obs}|}{|\mathcal{N}_i^{obs} \cup \mathcal{N}_j^{obs}|} , \tag{7.4}$$

and a variation on this idea due to Liben-Nowell and Kleinberg [1], extending a more general idea of Adamic and Adar [8], of the form

$$s(i, j) = \sum_{k \in \mathcal{N}_i^{obs} \cap \mathcal{N}_j^{obs}} \frac{1}{\log |\mathcal{N}_k^{obs}|} . \tag{7.5}$$

This last score in (7.5) has the effect of weighting more heavily those common neighbors of i and j that are themselves not highly connected.

---

[2]These plots combine a traditional boxplot (in the middle) with a kernel density estimate, running in a symmetric fashion to either side.

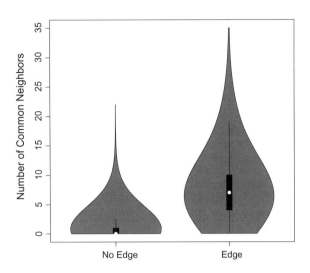

Fig. 7.2 Comparison of the number of common neighbors score statistic [i.e., (7.3)] in the network of French political blogs, grouped according to whether or not an edge is actually present between a vertex pair, for all vertex pairs

7.3 Association Network Inference

Often the rule used for defining edges in the network graph representation of a given data set is that there be a sufficient level of 'association' between certain attributes of the two incident vertices. Such *association networks* are found in many domains and include networks of citation patterns across scientific articles, networks of actors co-starring in movies, networks of regulatory influence among genes, and networks of functional connectivity between regions of the brain.

Frequently the rules defining edges in such networks are specified without statistics necessarily playing an explicit role. While such rule-based approaches are appropriate in some contexts, in other contexts, where issues of sampling or measurement error are of potential concern, it may be necessary to incorporate statistical principles and methods into the process of constructing an association network graph. That is, the problem becomes one of association network inference.

Formally, we may suppose we have a collection of elements represented as vertices $v \in V$. Furthermore, suppose that each such vertex v has corresponding to it a vector $\mathbf{x}$ of m observed vertex attributes, yielding a collection $\{\mathbf{x}_1, \ldots, \mathbf{x}_{N_v}\}$ of attribute vectors. Let $\mathsf{sim}(i, j)$ be a user-specified quantification of the inherent similarity between the pair of vertices $i, j \in V$, and assume that it is accompanied by a corresponding notion of what value(s) of $\mathsf{sim}(i, j)$ constitute a 'non-trivial' level of association between i and j. We are interested here in the case where sim is not itself directly observable, but nevertheless the attributes $\{\mathbf{x}_i\}$ contain sufficiently useful information to make inference on sim conceivable.

There are of course countless choices for sim in practice. Here we concentrate on two common and popular linear measures of association—correlation and partial correlation—and the corresponding methods for inferring an association network based upon them.

We will illustrate these methods in the context of gene regulatory networks. The data in `Ecoli.expr` contain two objects.

```
#7.4 1 > rm(list=ls())
     2 > data(Ecoli.data)
     3 > ls()
     4 [1] "Ecoli.expr" "regDB.adj"
```

The first is a 40 by 153 matrix of (log) gene expression levels in the bacteria *Escherichia coli (E. coli)*, measured for 153 genes under each of 40 different experimental conditions, averaged over three replicates of each experiment. The data are a subset of those published in [9]. The genes are actually so-called transcription factors, which in general are known to play a particularly important role in the network of gene regulatory relationships. The experiments were genetic perturbation experiments, in which a given gene was 'turned off', for each of 40 different genes. The gene expression measurements are an indication of the level of activity of each of the 153 genes under each of the given experimental perturbations.

In Fig. 7.3 is shown a heatmap visualization of these data.

```
#7.5 1 > heatmap(scale(Ecoli.expr), Rowv=NA)
```

The genes (columns) have been ordered according to a hierarchical clustering of their corresponding vectors of expression levels. Due to the nature of the process of gene regulation, the expression levels of gene regulatory pairs often can be expected to vary together. We see some visual evidence of such associations in the figure, where certain genes show similar behavior across certain subsets of experiments. This fact suggests the value of constructing an association network from these data, as a proxy for the underlying network(s) of gene regulatory relationships at work, a common practice in systems biology.

The second object in this data set is an adjacency matrix summarizing our (incomplete) knowledge of actual regulatory relationships in *E. coli*, extracted from the RegulonDB database[3] at the same time the experimental data were collected. We coerce this matrix into a network object

```
#7.6 1 > library(igraph)
     2 > g.regDB <- graph_from_adjacency_matrix(regDB.adj,
     3 +                                         "undirected")
     4 > summary(g.regDB)
     5 IGRAPH e14f679 UN-- 153 209 --
     6 + attr: name (v/c)
```

and note that there are 209 known regulatory pairs represented. Visualizing this graph

```
#7.7 1 > plot(g.regDB, vertex.size=3, vertex.label=NA)
```

we see, examining the result in Fig. 7.4, that these pairs are largely contained within a single connected component.

---

[3]http://regulondb.ccg.unam.mx/.

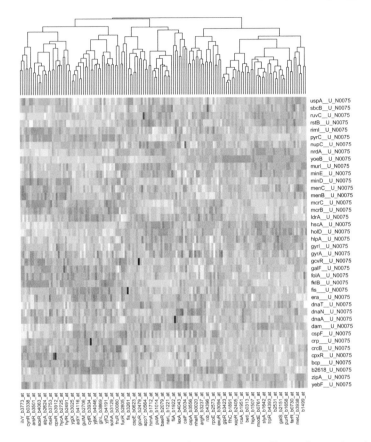

Fig. 7.3 Heatmap visualization of gene expression patterns over 40 experiments (*rows*) for 153 genes (*columns*)

Fig. 7.4 Network of known regulatory relationships among 153 genes in *E. coli*

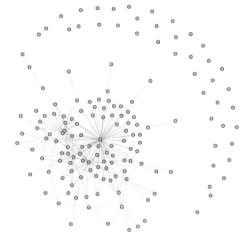

The information in the graph g.regDB represents a type of aggregate truth set, in that some or all of these gene–gene regulatory relationships may be active in the biological organism under a given set of conditions. As such, the graph will be useful as a point of reference in evaluating on these data the methods of association network inference we introduce below.

7.3.1 Correlation Networks

Let X be a (continuous) random variable of interest corresponding to the vertices in V. A standard measure of similarity between vertex pairs is $\text{sim}(i, j) = \rho_{ij}$, where

$$\rho_{ij} = \text{corr}_{X_i, X_j} = \frac{\sigma_{ij}}{\sqrt{\sigma_{ii}\sigma_{jj}}} \tag{7.6}$$

is the *Pearson product-moment correlation* between X_i and X_j, expressed in terms of the entries of the covariance matrix $\Sigma = \{\sigma_{ij}\}$ of the random vector $(X_1, \ldots, X_{N_v})^T$ of vertex attributes.

Given this choice of similarity, a natural criterion defining association between i and j is that ρ_{ij} be non-zero. The corresponding association graph G is then just the graph (V, E) with edge set

$$E = \left\{ \{i, j\} \in V^{(2)} : \rho_{ij} \neq 0 \right\} . \tag{7.7}$$

This graph is often called a *covariance (correlation) graph*.

Given a set of observations of the X_i, the task of inferring this association network graph can be taken as equivalent to that of inferring the set of non-zero correlations. One way this task can be approached is through testing[4] the hypotheses

$$H_0 : \rho_{ij} = 0 \quad \text{versus} \quad H_1 : \rho_{ij} \neq 0 . \tag{7.8}$$

However, there are at least three important issues to be faced in doing so. First, there is the choice of test statistic to be used. Second, given a test statistic, an appropriate null distribution must be determined for the evaluation of statistical significance. And third, there is the fact that a large number of tests are to be conducted simultaneously (i.e., for all $N_v(N_v - 1)/2$ potential edges), which implies that the problem of multiple testing must be addressed.

Suppose that for each vertex $i \in V$, we have n independent observations $x_{i1}, \ldots, x_{in}$ of X_i. As test statistics, the *empirical correlations*

---

[4]Another way is through the use of penalized regression methods, similar to those that we discuss in Sect. 7.3.3.

$$\hat{\rho}_{ij} = \frac{\hat{\sigma}_{ij}}{\sqrt{\hat{\sigma}_{ii}\hat{\sigma}_{jj}}} \tag{7.9}$$

are a common choice, where the $\hat{\sigma}_{ij}$ are the corresponding empirical versions of the σ_{ij} in (7.6).

To calculate these values across all gene pairs in our expression data is straight-forward.

```
#7.8 1 > mycorr <- cor(Ecoli.expr)
```

However, it is more common to work with transformations of these values. For example, Fisher's transformation,

$$z_{ij} = \tanh^{-1}(\hat{\rho}_{ij}) = \frac{1}{2}\log\left[\frac{(1 + \hat{\rho}_{ij})}{(1 - \hat{\rho}_{ij})}\right] \tag{7.10}$$

can be helpful as a variance stabilizing transformation. If the pair of variables (X_i, X_j) has a bivariate normal distribution, the density of z_{ij} under $H_0 : \rho_{ij} = 0$ is known to be well approximated by that of a Gaussian random variable with mean zero and variance $1/(n - 3)$, for sufficiently large n.

Accordingly, we transform the correlations in `mycorr` following (7.10),

```
#7.9 1 > z <- 0.5 * log((1 + mycorr) / (1 - mycorr))
```

and calculate p-values by comparing to the appropriate normal distribution.

```
#7.10 1 > z.vec <- z[upper.tri(z)]
     2 > n <- dim(Ecoli.expr)[1]
     3 > corr.pvals <- 2 * pnorm(abs(z.vec), 0,
     4 +    sqrt(1 / (n-3)), lower.tail=FALSE)
```

In assessing these p-values, however, it is necessary to account for the fact that we are conducting

```
#7.11 1 > length(corr.pvals)
     2 [1] 11628
```

tests simultaneously. The R function `p.adjust` may be used to calculate p-values adjusted for multiple testing. Here we apply a Benjamini-Hochberg adjustment, wherein p-values are adjusted through control of the false discovery rate.[5]

```
#7.12 1 > corr.pvals.adj <- p.adjust(corr.pvals, "BH")
```

---

[5]The *false discovery rate* (FDR) is defined to be

$$\text{FDR} = \mathbb{E}\left(\frac{R_{false}}{R} \,\middle|\, R > 0\right)\mathbb{P}(R > 0) \,, \tag{7.11}$$

where R is the number of rejections among m tests and R_{false} is the number of false rejections. Benjamini and Hochberg [10] provide a method for controlling the FDR at a user-specified level γ by rejecting the null hypothesis for all tests associated with p-values $p_{(j)} \leq (j/m)\gamma$, where $p_{(j)}$ is the jth smallest p-value among the m tests.

Comparing these values to a standard 0.05 significance level, we find a total of

```
#7.13  1 > length(corr.pvals.adj[corr.pvals.adj < 0.05])
       2 [1] 5227
```

gene pairs implicated. This number is far too large to be realistic, and an order of magnitude larger than in the aggregate network shown in Fig. 7.4. Simple diagnostics (e.g., using the qqnorm function to produce normal QQ plots) show that, unfortunately, the assumption of normality for the Fisher transformed correlations z is problematic with these data. As a result, any attempt to assign edge status to the 5227 'significant' vertex pairs above is highly suspect.

Fortunately, being in a decidedly 'data rich' situation, with over 11 thousand unique elements in z, we have the luxury to use data-dependent methods to learn the null distribution. The R package **fdrtool** implements several such methods.

```
#7.14  1 > library(fdrtool)
```

The methods in this package share the characteristic that they are all based on mixture models for some statistic S, of the form

$$f(s) = \eta_0 f_0(s) + (1 - \eta_0) f_1(s) , \qquad (7.12)$$

where $f_0(s)$ denotes the probability density function of S under the null hypothesis H_0, and $f_1(s)$, under the alternative hypothesis H_1, with η_0 effectively serving as the fraction of true null hypotheses expected in the data. Both f_0 and f_1 are estimated from the data, as is the mixing parameter η_0. In this framework, multiple testing is controlled through control of various notions of false discovery rates.

Here, taking the statistic S to be the empirical correlation (7.9) and using the default options in the main function fdrtool within **fdrtool**,

```
#7.15  1 > mycorr.vec <- mycorr[upper.tri(mycorr)]
       2 > fdr <- fdrtool(mycorr.vec, statistic="correlation")
```

results in the output summarized in Fig. 7.5. Note that the histogram of (absolute) empirical correlation coefficients shows a single and widely dispersed mode whose shape, although little resembling a normal distribution, can be easily estimated from the data. Note too, however, that with an estimate of $\hat{\eta}_0 = 1$, the method is indicating that all of the correlation coefficients are judged to come from the single density function f_0. Accordingly, the false discovery rate is estimated to be 100 % no matter what threshold might be applied to the correlations.

Hence, using an empirically derived null distribution, which more accurately captures the behavior of the data in this case, we are led to the conclusion that the correlation network we seek to infer is in fact the empty graph. A more apt interpretation, however, is simply that the combination of (a) choice of similarity, (b) sample size, and (c) distribution of effect size together have made it impossible in this data to discern a clear difference between edges and non-edges. We shall see momentarily that the use of a more nuanced notion of similarity leads to decidedly different results.

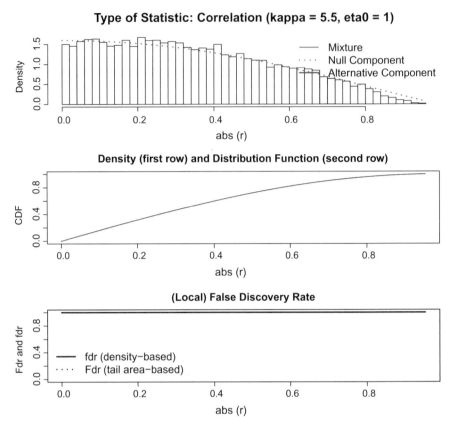

Fig. 7.5 Analysis of the empirical correlation coefficients for the gene expression data. *Top*: Estimated components f_0 and f_1 of the mixture in (7.12). *Middle*: Estimated density f. *Bottom*: False discovery rate

7.3.2 Partial Correlation Networks

The oft-cited dictum *'correlation does not imply causation'* should be kept in mind when constructing association networks based on Pearson's correlation and the like. Two vertices $i, j \in V$ may have highly correlated attributes X_i and X_j because the vertices somehow strongly 'influence' each other in a direct fashion. Alternatively, however, their correlation may be high primarily because, for example, they each are strongly influenced by a third vertex, say $k \in V$, and hence X_i and X_j are each highly correlated with X_k. The extent to which this issue is problematic or not will of course depend in no small part on the intended usage of the network graph G we seek to infer. But if it is felt desirable to construct a graph G where the inferred edges are more reflective of direct influence among vertices, rather than indirect influence, the notion of *partial correlation* becomes relevant.

In words, the partial correlation of attributes X_i and X_j of vertices $i, j \in V$, defined with respect to the attributes $X_{k_1}, \ldots, X_{k_m}$ of vertices $k_1, \ldots, k_m \in V \setminus \{i, j\}$, is the correlation between X_i and X_j left over after adjusting for those effects of $X_{k_1}, \ldots, X_{k_m}$ common to both. Formally, letting $S_m = \{k_1, \ldots, k_m\}$, we define the partial correlation of X_i and X_j, adjusting for $\mathbf{X}_{S_m} = (X_{k_1}, \ldots, X_{k_m})^T$, as

$$\rho_{ij|S_m} = \frac{\sigma_{ij|S_m}}{\sqrt{\sigma_{ii|S_m} \sigma_{jj|S_m}}} . \tag{7.13}$$

Here $\sigma_{ii|S_m}, \sigma_{jj|S_m}$, and $\sigma_{ij|S_m} = \sigma_{ji|S_m}$ are the diagonal and off-diagonal elements, respectively, of the 2×2 partial covariance matrix

$$\Sigma_{11|2} = \Sigma_{11} - \Sigma_{12} \Sigma_{22}^{-1} \Sigma_{21} , \tag{7.14}$$

where Σ_{11}, Σ_{22}, and $\Sigma_{12} = \Sigma_{21}^T$ are defined through the partitioned covariance matrix

$$\mathrm{Cov} \begin{pmatrix} \mathbf{W}_1 \\ \mathbf{W}_2 \end{pmatrix} = \begin{bmatrix} \Sigma_{11} & \Sigma_{12} \\ \Sigma_{21} & \Sigma_{22} \end{bmatrix} , \tag{7.15}$$

for $\mathbf{W}_1 = (X_i, X_j)^T$ and $\mathbf{W}_2 = \mathbf{X}_{S_m}$. If $(X_i, X_j, X_{k_1}, \ldots, X_{k_m})^T$ has a multivariate Gaussian distribution, then $\rho_{ij|S_m} = 0$ if and only if X_i and X_j are independent conditional on $\mathbf{X}_{S_m}$. For more general distributions, however, zero partial correlation will not necessarily imply independence (the converse, of course, is still true).

Note that in the case of $m = 0$, the partial correlation in (7.13) reduces to the Pearson correlation in (7.6). In addition, there are recursive expressions that relate each mth order partial correlation coefficient to three $(m - 1)$th order coefficients. For example, in the case of $m = 1$, the partial correlation of the attribute values X_i and X_j, for two vertices i and j, adjusted for the attribute value X_k of a third vertex k, is given by

$$\rho_{ij|k} = \frac{\rho_{ij} - \rho_{ik} \rho_{jk}}{\sqrt{\left(1 - \rho_{ik}^2\right) \left(1 - \rho_{jk}^2\right)}} . \tag{7.16}$$

For the general case see, for example, Anderson [11, Sect. 2.5.3].

Partial correlations can be used in various ways for defining an association network graph G, with respect to vertex attributes $X_1, \ldots, X_{N_v}$. For example, for a given choice of m, we may dictate that an edge be present only when there is correlation between X_i and X_j regardless of which m other vertices are conditioned upon. That is,

$$E = \left\{ \{i, j\} \in V^{(2)} : \rho_{ij|S_m} \neq 0, \text{ for all } S_m \in V_{\setminus \{i,j\}}^{(m)} \right\} , \tag{7.17}$$

where $V_{\setminus \{i,j\}}^{(m)}$ is the collection of all unordered subsets of m (distinct) vertices from $V \setminus \{i, j\}$. Other choices are clearly possible as well.

Under the definition of edges in (7.17), the problem of determining the presence or absence of a potential edge $\{i, j\}$ in G can be represented as one of testing

$$H_0 : \rho_{ij|S_m} = 0 \quad \text{for some} \quad S_m \in V_{\setminus\{i,j\}}^{(m)}$$

versus

$$H_1 : \rho_{ij|S_m} \neq 0 \quad \text{for all} \quad S_m \in V_{\setminus\{i,j\}}^{(m)} . \tag{7.18}$$

In order to infer an association network graph in this context, given measurements $x_{i1}, \ldots, x_{in}$ for each vertex $i \in V$, we must again select a test statistic, construct an appropriate null distribution, and adjust for multiple testing, as above.

Analogous to the empirical correlation coefficient $\hat{\rho}_{ij}$ in (7.9), the empirical partial correlation coefficient, $\hat{\rho}_{ij|S_m}$, is a natural estimate of $\rho_{ij|S_m}$. And, similarly, Fisher's transformation

$$z_{ij|S_m} = \frac{1}{2} \log \left[\frac{(1 + \hat{\rho}_{ij|S_m})}{(1 - \hat{\rho}_{ij|S_m})} \right] \tag{7.19}$$

is often used in place of the partial correlations themselves. Under the assumption that the joint distribution of attributes $(X_i, X_j, X_{k_1}, \ldots, X_{k_m})^T$ is a multivariate normal distribution, this statistic has an approximate normal distribution, with mean 0 and variance $1/(n - m - 3)$.

In approaching the testing problem in (7.18), it can be convenient to consider it as a collection of smaller testing sub-problems of the form

$$H_0' : \rho_{ij|S_m} = 0 \quad \text{versus} \quad H_1' : \rho_{ij|S_m} \neq 0 , \tag{7.20}$$

for which $\hat{\rho}_{ij|S_m}$ is a natural test statistic. A test of (7.18) may then be constructed from the tests of the sub-problems (7.20) through aggregation. For example, Wille et al. [12] suggest combining the information in p-values $p_{ij|S_m}$, over all S_m, by defining

$$p_{ij,\max} = \max \left\{ p_{ij|S_m} : S_m \in V_{\setminus\{i,j\}}^{(m)} \right\} \tag{7.21}$$

to be the p-value for the test of (7.18), where the $p_{ij|S_m}$ are the p-values for the tests of (7.20).

By way of illustration, recall the gene expression data Ecoli.expr that we analyzed in the previous section. We let $m = 1$, meaning that we define our network graph G that we will infer to be an association network where edges indicate correlation between the expression of gene pairs even after adjusting for the contributions of any other single gene. The following code utilizes the recursion in (7.16) to compute the corresponding empirical partial correlation coefficients, and approximates the distribution of the Fisher transformation of each such coefficient by a normal, with mean 0 and variance $1/(n - 4)$. Finally, p-values are assigned according to (7.21).

```
#7.16  1  > pcorr.pvals <- matrix(0, dim(mycorr)[1],
       2  +      dim(mycorr)[2])
       3  > for(i in seq(1, 153)){
       4  +    for(j in seq(1, 153)){
       5  +      rowi <- mycorr[i, -c(i, j)]
       6  +      rowj <- mycorr[j, -c(i, j)]
       7  +      tmp <- (mycorr[i, j] -
       8  +        rowi*rowj)/sqrt((1-rowi^2) * (1-rowj^2))
       9  +      tmp.zvals <- (0.5) * log((1+tmp) / (1-tmp))
      10  +      tmp.s.zvals <- sqrt(n-4) * tmp.zvals
      11  +      tmp.pvals <- 2 * pnorm(abs(tmp.s.zvals),
      12  +        0, 1, lower.tail=FALSE)
      13  +      pcorr.pvals[i, j] <- max(tmp.pvals)
      14  +    }
      15  + }
```

Adjusting for multiple testing, as in the previous section,

```
#7.17  1  > pcorr.pvals.vec <- pcorr.pvals[lower.tri(pcorr.pvals)]
       2  > pcorr.pvals.adj <- p.adjust(pcorr.pvals.vec, "BH")
```

and applying a nominal threshold of 0.05, we see that a total of

```
#7.18  1  > pcorr.edges <- (pcorr.pvals.adj < 0.05)
       2  > length(pcorr.pvals.adj[pcorr.edges])
       3  [1] 25
```

edges have been discovered under this analysis. Creating the corresponding network graph

```
#7.19  1  > pcorr.A <- matrix(0, 153, 153)
       2  > pcorr.A[lower.tri(pcorr.A)] <- as.numeric(pcorr.edges)
       3  > g.pcorr <- graph_from_adjacency_matrix(pcorr.A,
       4  +                                        "undirected")
```

and comparing to the aggregate network in Fig. 7.4,

```
#7.20  1  > intersection(g.regDB, g.pcorr, byname=FALSE)
       2  IGRAPH 7dd33fa UN-- 153 4 --
       3  + attr: name (v/c)
       4  + edges from 7dd33fa (vertex names):
       5  [1] yhiW_b3515_at--yhiX_b3516_at
       6  [2] rhaR_b3906_at--rhaS_b3905_at
       7  [3] marA_b1531_at--marR_b1530_at
       8  [4] gutM_b2706_at--srlR_b2707_at
```

we find that four of these 25 edges are among those established in the biological literature.

Alternatively, the analogous analysis of these same data using fdrtool, fitting a mixture model directly now to the p-values in pcorr.pvals.vec, yields the same number of edges.

```
#7.21  1  > fdr <- fdrtool(pcorr.pvals.vec, statistic="pvalue",
       2  +    plot=FALSE)
       3  > pcorr.edges.2 <- (fdr$qval < 0.05)
       4  > length(fdr$qval[pcorr.edges.2])
       5  [1] 25
```

It may be easily verified that these are in fact the same edges as above, thus suggesting a certain robustness of our results.

7.3.3 Gaussian Graphical Model Networks

A special—and, indeed, popular—case of the use of partial correlation coefficients is when $m = N_v - 2$ and the attributes are assumed to have a multivariate Gaussian joint distribution. Here the partial correlation between attributes of two vertices is defined conditional upon the attribute information at *all* other vertices. Denoting these coefficients as $\rho_{ij|V\setminus\{i,j\}}$, under the Gaussian assumption the vertices $i, j \in V$ have partial correlation $\rho_{ij|V\setminus\{i,j\}} = 0$ if and only if X_i and X_j are conditionally independent given all of the other attributes. The graph $G = (V, E)$ with edge set

$$E = \left\{ \{i, j\} \in V^{(2)} \; : \; \rho_{ij|V\setminus\{i,j\}} \neq 0 \right\} \tag{7.22}$$

is called a *conditional independence graph*.[6] The overall model, combining the multivariate Gaussian distribution with the graph G, is called a *Gaussian graphical model*.

A useful result in the context of Gaussian graphical models is that the partial correlation coefficients may be expressed in the form

$$\rho_{ij|V\setminus\{i,j\}} = \frac{-\omega_{ij}}{\sqrt{\omega_{ii}\omega_{jj}}} \; , \tag{7.23}$$

where ω_{ij} is the (i, j)th entry of $\Omega = \Sigma^{-1}$, the inverse of the covariance matrix Σ of the vector $(X_1, \ldots, X_{N_v})^T$ of vertex attributes. See Lauritzen [13, Sect. 5.1] or Whittaker [14, Sect. 5.8], for example. The matrix Ω is known as the *concentration* or *precision* matrix, and its non-zero off-diagonal entries, occurring as they do in the numerator of (7.23), are linked in one-to-one correspondence with the edges in G. As a result, G also is sometimes referred to as a *concentration graph*.

The problem of inferring G from data in this context was originally termed the 'covariance selection problem' by Dempster [15]). Traditionally, recursive, likelihood-based procedures have been used that effectively test the hypotheses

$$H_0 : \rho_{ij|V\setminus\{i,j\}} = 0 \quad \text{versus} \quad H_1 : \rho_{ij|V\setminus\{i,j\}} \neq 0 \; , \tag{7.24}$$

using the empirical partial correlations $\hat{\rho}_{ij|V\setminus\{i,j\}}$, defined as in (7.23), but with the entries of $\hat{\Omega} = \hat{\Sigma}^{-1}$ in place of those of Ω, where $\hat{\Sigma}$ is the usual unbiased estimate of covariance. However, for large-scale network graphs, it has arguably become the

---

[6]Note that it is actually the non-edges in G that correspond to conditional independence, and the edges, to conditional dependence.

standard to use penalized regression methods to infer G, exploiting a variation of the well-known connection between linear correlation and linear regression.

Suppose that the random vector $(X_1, \ldots, X_{N_v})^T$ of vertex attributes has a multivariate Gaussian distribution, with covariance Σ, as assumed above, and also zero mean. Then a standard result yields that, for the attribute X_i of a fixed vertex i, given the values of the attributes at the remaining vertices, say $\mathbf{X}^{(-i)} = (X_1, \ldots, X_{i-1}, X_{i+1}, \ldots, X_{N_v})^T$, its conditional expectation has the form

$$\mathbb{E}[\, X_i \mid \mathbf{X}^{(-i)} = \mathbf{x}^{(-i)}\,] = \left(\beta^{(-i)}\right)^T \mathbf{x}^{(-i)} , \tag{7.25}$$

where $\beta^{(-i)}$ is an $(N_v - 1)$-length parameter vector. See, for example, Johnson and Wichern [16, Chap. 4]. Furthermore, importantly, the entries of $\beta^{(-i)}$ can be expressed in terms of the entries of the precision matrix Ω, as $\beta_j^{(-i)} = -\omega_{ij}/\omega_{ii}$. See Lauritzen [13, Appendix C]. Hence, a potential edge $\{i, j\}$ is in the edge set E defined by (7.22) if and only if $\beta_j^{(-i)}$ (and, therefore, also $\beta_i^{(-j)}$) is not equal to zero.

These observations suggest that inference of G be pursued via inference of the non-zero elements of $\beta^{(-i)}$ in (7.25), using regression-based methods of estimation and variable selection. In fact, it can be shown that the vector $\beta^{(-i)}$ is the solution to the optimization problem

$$\arg \min_{\tilde{\beta}:\tilde{\beta}_i=0} \ \mathbb{E}\left[\left(X_i - \left(\tilde{\beta}^{(-i)}\right)^T \mathbf{X}^{(-i)} \right)^2 \right]. \tag{7.26}$$

It is natural, therefore, to replace this problem by a corresponding least-squares optimization. However, because there will be $N_v - 1$ variables in the regression for each X_i, and in addition it may be the case that $n \ll N_v$, a penalized regression strategy is prudent.

Meinshausen and Bühlmann [17], for example, have suggested using estimates of the form

$$\hat{\beta}^{(-i)} = \arg \min_{\beta:\beta_i=0} \sum_{k=1}^{n} \left(x_{ik} - \left(\beta^{(-i)}\right)^T \mathbf{x}_k^{(-i)} \right)^2 + \lambda \sum_{j \neq i} \left| \beta_j^{(-i)} \right|. \tag{7.27}$$

This strategy is based on the use of the Lasso method of Tibshirani [18], which not only estimates the coefficients in $\beta^{(-i)}$ but also forces values $\hat{\beta}_j^{(-i)} = 0$ where the association between X_i and X_j is felt to be too weak, with respect to the choice of penalty parameter λ. In other words, the Lasso methodology performs simultaneous estimation and variable selection, a characteristic due to the particular form of the penalty.

There are several practical concerns that must be addressed in order to put the above ideas into practice. These include (i) assessing the extent to which the data conform to a multivariate normal assumption, and (ii) choice of the penalty parameter λ. The R package **huge**—for 'high-dimensional, undirected graph estimation'—integrates solutions addressing both of these issues. The main function huge generates an initial set of estimates, following several pre-processing steps that seek to transform the data to have marginal distributions close to normal and to stabilize the overall estimation problem. The default method of estimation adopts the criterion in (7.27) above.

```
#7.22  1 > library(huge)
       2 > set.seed(42)
       3 > huge.out <- huge(Ecoli.expr)
```

The choice of penalty parameter can be determined by either of two procedures, the first being a so-called 'rotational information criterion' (i.e., in the spirit of more traditional criteria like AIC or BIC), and the second, a method based on principles of subsampling, called stability selection. The former procedure is understood to be prone to under-selection, while the latter, to over-selection.

Applying the former to our gene expression data returns an empty graph.

```
#7.23  1 > huge.opt <- huge.select(huge.out, criterion="ric")
       2 > g.huge <- graph_from_adjacency_matrix(huge.opt$refit,
       3 +                                        "undirected")
       4 > ecount(g.huge)
       5 [1] 0
```

Conversely, applying the latter, we arrive at a network graph of nontrivial connectivity.

```
#7.24  1 > huge.opt <- huge.select(huge.out, criterion="stars")
       2 > g.huge <- graph_from_adjacency_matrix(huge.opt$refit,
       3                                          "undirected")
       4 > summary(g.huge)
       5 IGRAPH 0f13a7e U--- 153 786 --
```

Comparisons indicate that this new network fully contains the network we obtained previously using partial correlation

```
#7.25  1 > intersection(g.pcorr, g.huge)
       2 IGRAPH 6948e99 U--- 153 25 --
       3 + edges from 6948e99:
       4 [1]   145--146 144--146 112--125 112--113 109--138
       5 [6] 108--135  97--111  96--119  92--107  87--104
       6 [11]  86-- 87  84--129  81--137  72--141  70-- 75
       7 [16]  60--127  46-- 77  42-- 43  38--153  37-- 98
       8 [21]  27--123  21-- 33  12--135   9-- 90   3-- 60
```

and, moreover, contains 21 edges among those established in the biological literature.

```
#7.26  1 > intersection(g.regDB, g.huge, byname=FALSE)
       2 IGRAPH 662d795 UN-- 153 21 --
       3 + attr: name (v/c)
       4 + edges from 662d795 (vertex names):
       5 [1] yhiW_b3515_at--yhiX_b3516_at
       6 [2] tdcA_b3118_at--tdcR_b3119_at
```

```
 7  [3] rpoE_b2573_at--rpoH_b3461_at
 8  [4] rpoD_b3067_at--tyrR_b1323_at
 9  [5] rhaR_b3906_at--rhaS_b3905_at
10  [6] nac_b1988_at --putA_b1014_at
11  [7] marA_b1531_at--marR_b1530_at
12  [8] malT_b3418_at--rpoD_b3067_at
13  + ... omitted several edges
```

7.4 Tomographic Network Topology Inference

Tomographic network topology inference is named in analogy to tomographic imaging[7] and refers to the inference of 'interior' components of a network—both vertices and edges—from data obtained at some subset of 'exterior' vertices. Here 'exterior' and 'interior' are somewhat relative terms and generally are used simply to distinguish between vertices where it is and is not possible (or, at least, not convenient) to obtain measurements. For example, in computer networks, desktop and laptop computers are typical instances of 'exterior' vertices, while Internet routers to which we do not have access are effectively 'interior' vertices.

With information only obtainable through measurements at 'exterior' vertices, the tomographic network topology inference problem can be expected to be quite difficult in general. This difficulty is primarily due to the fact that, for a given set of measurements, there are likely many network topologies that conceivably could have generated them, and without any further constraints on aspects like the number of internal vertices and edges, and the manner in which they connect to one another, we have no sensible way of choosing among these possible solutions.

In order to obtain useful solutions to such problems, therefore, additional information must be incorporated into the problem statement by way of model assumptions on the form of the internal structure to be inferred. A key structural simplification has been the restriction to inference of networks in the form of trees. In fact, nearly all work in this area to date has focused on this particular version of the problem.

7.4.1 Constraining the Problem: Tree Topologies

Recall that an (undirected) *tree* $T = (V_T, E_T)$ is a connected graph with no cycles. A *rooted tree* is a tree in which a vertex $r \in V_T$ has been singled out. The subset of vertices $R \subset V_T$ of degree one are called the *leaves*; we will refer to the vertices in $V \backslash \{\{r\} \cup R\}$ as the *internal vertices*. The edges in E_T are often referred to as *branches*.

---

[7]In the field of imaging, *tomography* refers to the process of imaging by sections, such as is done in medical imaging contexts like X-ray tomography or positron emission tomography (PET).

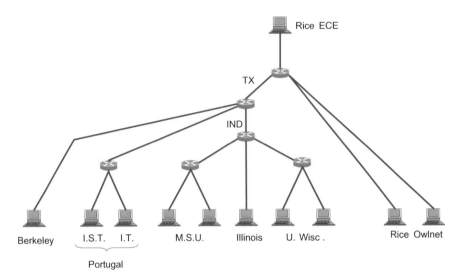

Fig. 7.6 Computer network topology from the experiment of Castro et al. [19], representing 'groundtruth'

For example, in Fig. 7.6 is shown the logical topology, during a period of 2001, of that portion of the Internet leading from a desktop computer in the Electrical and Computer Engineering Department at Rice University to similar machines at ten other university locations. Specifically, two of the destination machines were located on a different sub-network at Rice, two at separate locations in Portugal, and six at four other universities in the United States. The original machine at Rice forms the root of this tree, in this representation, while the other ten machines form the leaves. These machines are 'exterior' in the sense that they would be known to, for example, their users. The other vertices, as well as the branches, are all 'internal' in the sense that these typically would not be known to a standard computer user. Hence, the learning of this internal structure (from appropriate measurements) is of natural interest.

It is quite common to restrict attention to binary trees. A *binary tree* is a rooted tree in which, in moving from the root towards the leaves, each internal vertex has at most two children. Trees with more general branching structure (such as that in Fig. 7.6) can always be represented as binary trees.

We define our tomographic network inference problem as follows. Suppose that for a set of N_l vertices—possibly a subset of the vertices V of an arbitrary graph $G = (V, E)$—we have n independent and identically distributed observations of some random variables $\{X_1, \ldots, X_{N_l}\}$. Under the assumption that these vertices can be meaningfully identified with the leaves R of a tree T, we aim to find that tree $\hat{T}$ in the set $\mathscr{T}_{N_l}$ of all binary trees with N_l labeled leaves that best explains the data, in some well-defined sense. If we have knowledge of a root r, then the roots of the trees in $\mathscr{T}_{N_l}$ will all be identified with r. In some contexts we may also be interested in inferring a set of weights for the branches in $\hat{T}$.

There are two main areas in which work on this type of network inference problem may be found. The first is in biology and, in particular, the inference of phylogenies. Phylogenetic inference is concerned with the construction of trees (i.e., phylogenies) from data, for the purpose of describing evolutionary relationships among biological species. See, for example, the book by Felsenstein [20]; also useful, and shorter, are the two surveys by Holmes [21, 22]. The second is in computer network analysis and the identification of logical topologies (i.e., as opposed to physical topologies). In computer network topology identification, the goal is to infer the tree formed by a set of paths along which traffic flows from a given origin Internet address to a set of destination addresses. Castro et al. [23, Chap. 4] provide a survey of this area.

The tree shown in Fig. 7.6 is an example of such a computer network topology. In order to learn such topologies (since, in general, a map of the Internet over such scales is not available), various techniques for probing the Internet have been developed. These probes, typically in the form of tiny packets of information, are sent from a source computer (e.g., the root machine at Rice University in Fig. 7.6) to various destinations (e.g., the leaf machines in Fig. 7.6), and information reflective of the underlying topology is measured.

For example, Coates et al. [24] propose a measurement scheme they call 'sandwich probing,' which sends a sequence of three packets, two small and one large. The large packet is sent second, after the first small packet and before the second small packet. The two small packets are sent to one of a pair $\{i, j\}$ of leaf vertices, say i, while the large packet is sent to the other, that is, j. The basic idea is that the large packet induces a greater delay in the arrival of the second small packet at its destination, compared to that of the first, and that the difference in delays of the two small packets will vary in a manner reflective of how much of the paths are shared from the origin to i and j.

The data set `sandwichprobe` contains the results of this experiment, consisting of (i) the measured delay between small packet pairs (in microseconds), indexed by the destination to which the small and large packets were sent, respectively, and (ii) the identifiers of destinations (called 'host' machines).

```
#7.27  1 > data(sandwichprobe)
       2 > delaydata[1:5, ]
       3   DelayDiff SmallPktDest BigPktDest
       4 1       757            3         10
       5 2       608            6          2
       6 3       242            8          9
       7 4        84            1          8
       8 5      1000            7          3
       9 > host.locs
      10 [1] "IST"    "IT"    "UCBkly" "MSU1"   "MSU2"
      11 [6] "UIUC"   "UW1"   "UW2"    "Rice1"  "Rice2"
```

Fig. 7.7 shows an image representation of the mean delay differences, symmetrized to eliminate variations within each pair due to receipt of the small packets versus the large packet.

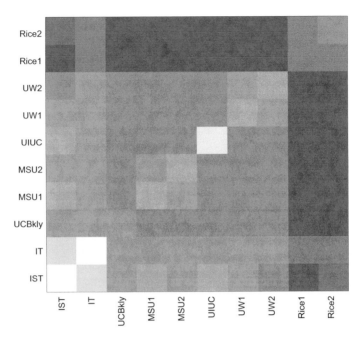

Fig. 7.7 Image representation of pairwise delay differences in the data of Coates et al. [24], with increasingly *darker red* corresponding to lower values, and increasingly *brighter yellow*, to higher values

```
#7.28  1  > meanmat <- with(delaydata, by(DelayDiff,
       2  +    list(SmallPktDest, BigPktDest), mean))
       3  > image(log(meanmat + t(meanmat)), xaxt="n", yaxt="n",
       4  +    col=heat.colors(16))
       5  > mtext(side=1, text=host.locs,
       6  +    at=seq(0.0,1.0,0.11), las=3)
       7  > mtext(side=2, text=host.locs,
       8  +    at=seq(0.0,1.0,0.11), las=1)
```

The hierarchical relationship among the destinations in the logical topology in Fig. 7.6 is clearly evident in the relative magnitudes of the mean delay differences. This association can be used to infer the topology from the delay differences.

7.4.2 Tomographic Inference of Tree Topologies: An Illustration

Broadly speaking, there are two classes of methods that have been developed quite extensively for the tomographic inference of tree topologies: (i) those based on hierarchical clustering and related ideas, and (ii) likelihood-based methods. Details associated with examples of the latter class of methods, being model-based, tend to be quite context-dependent. In contrast, simpler versions of the former class of methods are relatively straightforward to describe and may be implemented using existing tools in the R base package. Therefore, although a treatment of likelihood-based methods is beyond the scope of this book,[8] we will take a look at a simple example of how hierarchical clustering may be applied to this problem.

Given a set of N_l objects, recall from our discussion in Sect. 4.4.1 that hierarchical clustering is concerned with the grouping of those objects into hierarchically nested sets of partitions, based on some notion of their (dis)similarity. And, in particular, recall that generally these nested partitions are represented using a tree. The usage of hierarchical clustering we saw in Sect. 4.4.1 was for the purpose of graph clustering. But it also is a natural tool to use for the tomographic inference of tree topologies, where we treat the N_l leaves as the 'objects' to be clustered and the tree correspond-ing to the resulting clustering as our inferred tree $\hat{T}$. It is perhaps worth emphasizing that, whereas in standard applications of hierarchical clustering the goal often ulti-mately is to select just one of the nested partitions to represent the data, here the tree corresponding to the entire set of partitions is our focus.

Recall that hierarchical clustering methods require some notion of (dis)similarity, which is usually constructed from the observed data, but sometimes measured directly. In the context of our tomographic inference problem, it is logical to use the n observations of the random variables $\{X_1, \ldots, X_{N_l}\}$ at the leaves to derive an $N_l \times N_l$ matrix of (dis)similarities.

For example, the information in the matrix delaydata may be used to compute the (squared) Euclidean distance between delays at each pair of destinations.

```
#7.29  1 > SSDelayDiff <- with(delaydata, by(DelayDiff^2,
       2 +    list(SmallPktDest, BigPktDest), sum))
```

The R function hclust may then be used for hierarchical clustering.

```
#7.30  1 > x <- as.dist(1 / sqrt(SSDelayDiff))
       2 > myclust <- hclust(x, method="average")
```

Here we have used the inverse of distance between delays as a similarity matrix. The choice of average for the hierarchical clustering method corresponds to using the unweighted pair group method with arithmetic mean (UPGMA), a standard choice and one that is quite similar to the more application-specific agglomerative likelihood tree (ALT) method of Coates et al. [24].

The result of our hierarchical clustering is displayed in Fig. 7.8, in the form of a dendrogram.

```
#7.31  1 > plot(myclust, labels=host.locs, axes=FALSE,
       2 +    ylab=NULL, ann=FALSE)
```

---

[8]A brief treatment of such methods, largely from the perspective of computer network topology inference, may be found in [25, Sect. 7.4]. In terms of software, there are a number of fairly com-prehensive packages in R dedicated to phylogenetic tree inference. See, for example, the packages **ape** and **phangorn**.

Fig. 7.8 Inferred topology, based on 'sandwich' probe measurements, using a hierarchical clustering algorithm

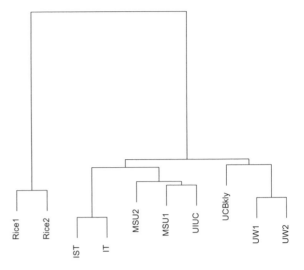

Comparing our inferred topology to the ground truth in Fig. 7.6, we see that many aspects of the latter are capture by the former. For example, the fact that the two destinations at Rice University are, naturally, much closer to the root machine at Rice than any of the other destinations is strongly indicated. Similarly, the two machines in Portugal are found grouped together, as are the two machines at the University of Wisconsin, and the three other machines in the midwest (i.e., two at Michigan State, and one at the University of Illinois). Curiously, however, the machine at Berkeley is inferred to be close to those at Wisconsin, although in reality they appear to be well-separated in the actual topology.

7.5 Additional Reading

Surveys of methods of link prediction include that of Lu and Zhou [26], in the physics literature, and Al Hasan and Zaki [27] or Martinez and colleagues [28], in the computer science literature. For the specific problem of inferring gene regulatory networks, there are many resources to be found, such as the survey of Noor et al. [29] or the recent chapter by Va and Sanguinetti [30]. Finally, for more information on tomographic inference of tree topologies, the literature in phylogenetics is extensive. See, for example, the book by Felsenstein [20] or the surveys by Holmes [21, 22].

References

1. D. Liben-Nowell and J. Kleinberg, "The link prediction problem for social networks," in *Proceedings of the 12th International Conference on Information and Knowledge Management*, 2003.
2. A. Popescul and L. Ungar, "Statistical relational learning for link prediction," in *Proceedings of the Workshop on Learning Statistical Models from Relational Data at IJCAI-2003*, 2003.
3. B. Taskar, M. Wong, P. Abbeel, and D. Koller, "Link prediction in relational data," *Advances in Neural Information Processing Systems*, vol. 16, 2004.
4. P. Hoff, "Multiplicative latent factor models for description and prediction of social networks," *Computational and Mathematical Organization Theory*, 2007.
5. D. Goldberg and F. Roth, "Assessing experimentally derived interactions in a small world," *Proceedings of the National Academy of Sciences*, vol. 100, no. 8, pp. 4372–4376, 2003.
6. J. Bader, A. Chaudhuri, J. Rothberg, and J. Chant, "Gaining confidence in high-throughput protein interaction networks," *Nature Biotechnology*, vol. 22, no. 1, pp. 78–85, 2004.
7. R. Little and D. Rubin, *Statistical Analysis with Missing Data, Second Edition*.New York: Wiley & Sons, Inc., 2002.
8. L. Adamic and E. Adar, "Friends and neighbors on the Web," *Social Networks*, vol. 25, no. 3, pp. 211–230, 2003.
9. J. Faith, B. Hayete, J. Thaden, I. Mogno, J. Wierzbowski, G. Cottarel, S. Kasif, J. Collins, and T. Gardner, "Large-scale mapping and validation of *Escherichia coli* transcriptional regulation from a compendium of expression profiles," *PLoS Biology*, vol. 5, no. 1, p. e8, 2007.
10. Y. Benjamini and Y. Hochberg, "Controlling the false discovery rate: a practical and powerful approach to multiple testing," *Journal of the Royal Statistical Society, Series B*, vol. 57, no. 1, pp. 289–300, 1995.
11. T. Anderson, *An Introduction to Multivariate Statistical Analysis, Second Edition*. New York: John Wiley & Sons, Inc., 1984.
12. A. Wille, P. Zimmermann, E. Vránova, A. Fürholz, O. Laule, S. Bleuler, L. Hennig, A. Prelić, P. Rohr, L. Thiele *et al.*, "Sparse graphical Gaussian modeling of the isoprenoid gene network in arabidopsis thaliana," *Genome Biology*, vol. 5, no. 11, p. R92, 2004.
13. S. Lauritzen, *Graphical Models*. Oxford: Oxford University Press, 1996.
14. J. Whittaker, *Graphical Models in Applied Multivariate Statistics*. Chichester: Wiley & Sons, 1990.
15. A. Dempster, "Covariance selection," *Biometrics*, vol. 28, no. 1, pp. 157–175, 1972.
16. R. Johnson and D. Wichern, *Applied Multivariate Statistical Analysis, Fifth Edition*. Upper Saddle River, NJ: Pearson Eduction, 2001.
17. N. Meinshausen and P. Bühlmann, "High-dimensional graphs and variable selection with the Lasso," *Annals of Statistics*, vol. 34, no. 3, pp. 1436–1462, 2006.
18. R. Tibshirani, "Regression shrinkage and selection via the lasso," *Journal of the Royal Statistical Society, Series B*, vol. 58, pp. 267–288, 1996.
19. R. Castro, M. Coates, and R. Nowak, "Likelihood-based hierarchical clustering," *IEEE Transactions on Signal Processing*, vol. 52, no. 8, pp. 2308–2321, 2004.
20. J. Felsenstein, *Inferring Phylogenies*. Sunderland, MA: Sinear Associates, 2004.
21. S. Holmes, "Phylogenies: An overview," in *Statistics and Genetics*, ser. IMA, E. Halloran and S. Geisser, Eds. New York: Springer-Verlag, 1999, vol. 81.
22. ——, "Statistics for phylogenetic trees," *Theoretical Population Biology*, vol. 63, pp. 17–32, 2003.
23. R. Castro, M. Coates, G. Liang, R. Nowak, and B. Yu, "Network tomography: recent developments," *Statistical Science*, vol. 19, no. 3, pp. 499–517, 2004.
24. M. Coates, R. Castro, R. Nowak, M. Gadhiok, R. King, and Y. Tsang, "Maximum likelihood network topology identification from edge-based unicast measurements," *Proceedings of the 2002 ACM SIGMETRICS International Conference on Measurement and Modeling of Computer Systems*, pp. 11–20, 2002.

25. E. Kolaczyk, *Statistical Analysis of Network Data: Methods and Models*. Springer Verlag, 2009.
26. L. Lü and T. Zhou, "Link prediction in complex networks: A survey," *Physica A: Statistical Mechanics and its Applications*, vol. 390, no. 6, pp. 1150–1170, 2011.
27. M. Al Hasan and M. J. Zaki, "A survey of link prediction in social networks," in *Social Network Data Analytics*. Springer, 2011, pp. 243–275.
28. V. Martínez, F. Berzal, and J.-C. Cubero, "A survey of link prediction in social networks," in *ACM Computing Surveys (CSUR)*. ACM, vol. 49, no. 4, pp. 69, 2017.
29. A. Noor, E. Serpedin, M. Nounou, H. Nounou, N. Mohamed, and L. Chouchane, "An overview of the statistical methods used for inferring gene regulatory networks and protein-protein interaction networks," *Advances in Bioinformatics*, vol. 2013, 2013.
30. G. Sanguinetti *et al.*, "Gene regulatory network inference: an introductory survey," in *Gene Regulatory Networks*. Springer, 2019, pp. 1–23.

Chapter 8
Modeling and Prediction for Processes on Network Graphs

8.1 Introduction

Throughout this book so far, we have seen numerous examples of network graphs that provide representations—useful for various purposes—of the interaction among elements in a system under study. Often, however, it is some quantity (or attribute) associated with each of the elements that ultimately is of most interest. In such settings it frequently is not unreasonable to expect that this quantity be influenced in an important manner by the interactions among the elements. For example, the behaviors and beliefs of people can be strongly influenced by their social interactions; proteins that are more similar to each other, with respect to their DNA sequence information, often are responsible for the same or related functional roles in a cell; computers more easily accessible to a computer infected with a virus may in turn themselves become more quickly infected; and the relative concentration of species in an environment (e.g., animal species in a forest or chemical species in a vat) can vary over time as a result of the nature of the relationships among species.

Quantities associated with such phenomena can usefully be thought of as stochastic processes defined on network graphs. More formally, they can be represented in terms of collections of random variables, say X, indexed on a network graph $G = (V, E)$, either of the form $\{X_i\}$, for $i \in V$, or $\{X_i(t)\}$, with t varying in a discrete or continuous manner over a range of times. For example, the functionality of proteins can be viewed as categorical variables associated with each $i \in V$. So too can various behaviors and beliefs of individuals in a social community be represented using such variables, possibly indexed in time. Similarly, the spread of a computer virus can be captured using a set of binary variables (i.e., 'infected' or 'not infected') that evolve over time.

We will refer to processes $\{X_i\}$ as *static processes* and $\{X_i(t)\}$ as *dynamic processes*. Given appropriate measurements, statistical tasks that arise in the study of such processes include their modeling and the inference of model parameters, and also, in particular, prediction. To date, most such work arguably has been done in the context of static processes, although this situation is changing.

© Springer Nature Switzerland AG 2020

E. D. Kolaczyk and G. Csárdi, *Statistical Analysis of Network Data with R*, Use R!,

https://doi.org/10.1007/978-3-030-44129-6_8

Accordingly, our presentation in this chapter will concentrate largely on the case of static processes, with the development of methods and models in Sect. 8.2 through 8.4. In addition, we briefly discuss the modeling and prediction of dynamic processes in Sect. 8.5. Some further discussion related to this latter topic may be found in Sect. 11.5.

8.2   Nearest Neighbor Methods

We begin by focusing on the problem of predicting a static process on a graph and demonstrate the feasibility of doing so by examining one of the simplest of methods for this task—nearest-neighbor prediction.

Consider a collection of vertex attributes, which, as in previous chapters, we will express succinctly in vector form as $\mathbf{X} = (X_i)$. Such attributes may be inherently independent of time, and hence form a truly static process, or perhaps more commonly, may constitute a 'snapshot' of a dynamic process in a given 'slice' of time. In Chaps. 6 and 7, such attributes were used in modeling and predicting the presence or absence of edges in a network graph G. That is, we modeled the behavior of the variables $\mathbf{Y} = [Y_{ij}]$, conditional on $\mathbf{X}$. Alternatively, however, in some contexts it may be the behavior of $\mathbf{X}$, conditional on $\mathbf{Y}$, that is of interest instead.

We illustrate through the problem of protein function prediction. Recall, from our discussion of validation of graph partitioning in Sect. 4.4.3 and of assortativity and mixing in Sect. 4.5, that the affinity of proteins to physically bind to each other is known to be directly related to their participation in common cellular functions. And, in fact, we found in our analysis of the data set `yeast` that the external assignment of proteins to functional classes correlated to a reasonable extent with their assignment to 'communities' by our graph partitioning algorithms. Furthermore, we saw examples of strong assortative mixing of protein function in the underlying network of protein–protein interactions. While the gold standard for establishing the functionality of proteins is through direct experimental validation (or 'assays'), results like these have been taken to suggest that, given the vast number of proteins yet to be annotated, it is natural to approach this problem from the perspective of statistical prediction. Approaches to protein function prediction that incorporate network-based information are standard.

The data set `ppiCC`

```
#8.1 1 > set.seed(42)
     2 > library(sand)
     3 > data(ppi.CC)
```

contains a network data object, called `ppi.CC`, that consists of a network of 241 interactions among 134 proteins, as well as various vertex attributes.

```
#8.2  1  > summary(ppi.CC)
      2  IGRAPH 2ce6d08 UN-- 134 241 --
      3  + attr: name (v/c), ICSC (v/n),
      4  | IPR000198 (v/n), IPR000403 (v/n),
      5  | IPR001806 (v/n), IPR001849 (v/n),
      6  | IPR002041 (v/n), IPR003527 (v/n)
```

These data pertain to Baker's yeast—the organism formally known as *Saccharomyces cerevisiae*. They were assembled by Jiang et al. [12], from various sources, and pertain to only those proteins annotated, as of January 2007, with the term *cell communication* in the gene ontology[1] (GO) database—a standard database for terms describing protein function. The vertex attribute ICSC is a binary vector

```
#8.3  1  > V(ppi.CC)$ICSC[1:10]
      2   [1] 1 1 1 1 1 0 1 1 1 1
```

indicating those proteins annotated with the GO term *intracellular signaling cascade* (ICSC), a specific form of cellular communication.

A visualization of this network is shown in Fig. 8.1.

```
#8.4  1  > V(ppi.CC)[ICSC == 1]$color <- "yellow"
      2  > V(ppi.CC)[ICSC == 0]$color <- "blue"
      3  > plot(ppi.CC, vertex.size=5, vertex.label=NA)
```

We can see that there is a good deal of homogeneity to the vertex labels, with neighbors frequently sharing the same color. This observation suggests that local prediction of ICSC on this network graph should be feasible.

A simple, but often quite effective, method for producing local predictions is the *nearest-neighbor* method. See Hastie et al. [10, Sect. 2.3.2], for example, for general background on nearest-neighbor methods. For networks, the nearest-neighbor method centers on the calculation, for a given vertex $i \in V$, of the nearest-neighbor average

$$\frac{\sum_{j \in \mathcal{N}_i} x_j}{|\mathcal{N}_i|} , \qquad (8.1)$$

i.e., the average of the values of the vertex attribute vector $\mathbf{X}$ in the neighborhood $\mathcal{N}_i$ of i. Here $|\mathcal{N}_i|$ denotes the number of neighbors of i in G. Calculation of these averages over all vertices $i \in V$ corresponds to a nearest-neighbor smoothing of $\mathbf{X}$ across the network.

In the context of protein function prediction, $\mathbf{X}$ is a binary vector, with entries indicating whether or not each protein is or is not annotated with a function of interest (e.g., ICSC). In predicting binary vertex attributes, the nearest-neighbor averages (8.1) typically are compared to some threshold. For example, a threshold of 0.5

---

[1]http://www.geneontology.org.

Fig. 8.1 Network of
interactions among proteins
known to be responsible for
cell communication in yeast.
Yellow vertices denote
proteins that are known to be
involved in intracellular
signaling cascades, a specific
form of communication in
the cell. The remaining
proteins are indicated in *blue*

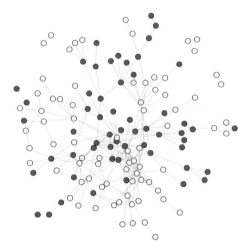

is commonly used, with a nearest-neighbor average greater than this value meaning
that a majority of neighbors have the characteristic indicated by $X = 1$, resulting in a
prediction for X_i of 1 as well. Such methods are also known as 'guilt-by-association'
methods in some fields.

In order to obtain some sense as to how effective the nearest-neighbor method
might be in predicting ICSC in our yeast data set, utilizing the information available
through protein–protein interactions, we can calculate the nearest-neighbor average
for each of the proteins in the giant connected component of our network.

```
#8.5  1 > clu <- components(ppi.CC)
      2 > ppi.CC.gc <- induced_subgraph(ppi.CC,
      3 +    clu$membership==which.max(clu$csize))
      4 > nn.ave <- sapply(V(ppi.CC.gc),
      5 +    function(x) mean(V(ppi.CC.gc)[nei(x)]$ICSC))
```

We then plot histograms of the resulting values, separated according to the status
of the vertex defining each neighborhood, i.e., according to the status of the 'ego'
vertex, in the terminology of social networks.

```
#8.6  1 > par(mfrow=c(2,1))
      2 > hist(nn.ave[V(ppi.CC.gc)$ICSC == 1], col="yellow",
      3 +    ylim=c(0, 30), xlab="Proportion Neighbors w/ ICSC",
      4 +    main="Egos w/ ICSC")
      5 > hist(nn.ave[V(ppi.CC.gc)$ICSC == 0], col="blue",
      6 +    ylim=c(0, 30), xlab="Proportion Neighbors w/ ICSC",
      7 +    main="Egos w/out ICSC")
```

The results, shown in Fig. 8.2, confirm that ICSC can be predicted with fairly good
accuracy.[2] In particular, using a threshold of 0.5 would yield an error rate of roughly
25%.

---

[2] A more rigorous assessment of predictive performance would use some version of cross-validation,
similar to what was used in Sect. 6.4. However, for the purposes of illustration, here and elsewhere

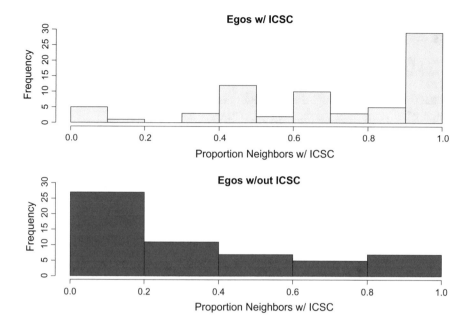

Fig. 8.2 Histograms of nearest-neighbor averages for the network shown in Fig. 8.1, separated according to the status of the vertex defining each neighborhood (i.e., 'ego')

```
#8.7 1 > nn.pred <- as.numeric(nn.ave > 0.5)
     2 > mean(as.numeric(nn.pred != V(ppi.CC.gc)$ICSC))
     3 [1] 0.2598425
```

Interestingly, we can push this illustration a bit further by taking advantage of the evolving nature of biological databases like GO. In particular, the proteins annotated in GO as not having a given biological function include both (1) those that indeed are *known* not to have that function, and (2) those whose status is simply unknown. As a result, by comparing against more recent versions of GO, it is sometimes possible to identify proteins whose status has changed for a given functional annotation, indicating that in the interim it has been discovered to in fact have that function.

The R package **GOstats**, a part of the **Bioconductor** package, may be used to manipulate and analyze Gene Ontology, as contained in the database GO.db.

```
#8.8 1 > if (!requireNamespace("BiocManager", quietly = TRUE))
     2 +     install.packages("BiocManager")
     3 > BiocManager::install()
     4 > BiocManager::install(c("GOstats","GO.db"))
     5 > library(GOstats)
     6 > library(GO.db)
```

---

in this chapter we instead content ourselves with examination of the fitted values produced by the various methods discussed, which, for sufficiently low to moderate numbers of predictions to be made, should be reflective of actual predictive performance.

And the annotations specific to the organism yeast can be obtained from the `org.Sc.sgd.db` database.

```
#8.9 1 > BiocManager::install("org.Sc.sgd.db")
     2 > library(org.Sc.sgd.db)
```

At the time of this writing, these annotations were last updated in January of 2017, roughly ten years after the data in `ppi.CC` were assembled.

We extract those proteins with the function ICSC—now subsumed[3] under the term intercellular signaling transduction (ICST), or GO label 003556—and keep only those that have been identified from direct experimental assay, as indicated by the evidence code 'IDA'.

```
#8.10 1 > x <- as.list(org.Sc.sgdGO2ALLORFS)
      2 > current.icst <- x[names(x) == "GO:0035556"]
      3 > ev.code <- names(current.icst[[1]])
      4 > icst.ida <- current.icst[[1]][ev.code == "IDA"]
```

We then separate out the names of those proteins that had ICSC in our original data

```
#8.11 1 > orig.icsc <- V(ppi.CC.gc)[ICSC == 1]$name
```

and similarly extract the names of those proteins under the new annotations that were present in the giant connected component of our original network.

```
#8.12 1 > candidates <- intersect(icst.ida, V(ppi.CC.gc)$name)
```

Among these candidates, there are seven that have been newly discovered to have ICSC, with the following names.

```
#8.13 1 > new.icsc <- setdiff(candidates, orig.icsc)
      2 > new.icsc
      3 [1] "YDL159W" "YDL235C" "YHL007C" "YIL033C"
      4 [5] "YIL147C" "YLR006C" "YLR362W"
```

And among these seven, we find that three of them would have been correctly predicted by comparing the value of their nearest-neighbor averages to a threshold of 0.5.

```
#8.14 1 > nn.ave[V(ppi.CC.gc)$name %in% new.icsc]
      2    YIL033C    YLR362W    YDL159W    YLR006C    YHL007C
      3 0.7500000 0.4166667 0.3333333 0.6666667 0.8750000
      4    YDL235C    YIL147C
      5 0.0000000 0.0000000
```

---

[3]Not only do the protein annotations in GO evolve over time, as biological knowledge continues to be developed and refined, but so too (to a lesser extent) do the actual names of those annotations, as is the case here, with intercellular signaling cascade being replaced in the GO database by intercellular signaling transduction.

8.3 Markov Random Fields

The principles underlying the nearest-neighbor method can be formalized and extended through the construction of appropriate statistical models. Such modeling can allow, for example, for probabilistically rigorous predictive statements as well as estimation and testing of model parameters. In addition, a modeling perspective facilitates a systematic approach to the inclusion of both network (endogenous) and non-network (exogenous) effects. It can also facilitate the handling of missing data. Markov random fields (MRFs) represent one well-developed modeling paradigm that achieves all of these goals.

8.3.1 General Characterization

Let $G = (V, E)$ be a graph and $\mathbf{X} = (X_1, \ldots, X_{N_v})^T$ be a collection of discrete random variables defined on V. We say that $\mathbf{X}$ is a *Markov random field* (MRF) on G if

$$\mathbb{P}\left(\mathbf{X} = \mathbf{x}\right) > 0, \quad \text{for all possible outcomes } \mathbf{x}, \tag{8.2}$$

and

$$\mathbb{P}\left(X_i = x_i \mid \mathbf{X}_{(-i)} = \mathbf{x}_{(-i)}\right) = \mathbb{P}\left(X_i = x_i \mid \mathbf{X}_{\mathcal{N}_i} = \mathbf{x}_{\mathcal{N}_i}\right), \tag{8.3}$$

where $\mathbf{X}_{(-i)}$ is the vector $(X_1, \ldots, X_{i-1}, X_{i+1}, \ldots, X_{N_v})^T$ and $\mathbf{X}_{\mathcal{N}_i}$ is the vector of all X_j for $j \in \mathcal{N}_i$. The positivity assumed in (8.2) for the joint distribution of $\mathbf{X}$ is simply a useful technical condition. The expression in (8.3) is the key Markov condition, asserting that X_i is conditionally independent of all other X_k, given the values of its neighbors, where the neighborhood structure is determined by G.

The concept of an MRF can be seen as a generalization of a Markov chain (common in the modeling of temporal data) and has its roots in statistical mechanics, going back to the work of Ising [11] on ferromagnetic fields. MRFs are used extensively in spatial statistics (e.g., see Cressie [6, Sect. 6.4]) and in image analysis (e.g., see Li [15]).

A key feature facilitating the practical usage of Markov random fields is their equivalence, under appropriate conditions, with *Gibbs random fields* i.e., random vectors $\mathbf{X}$ with distributions of the form

$$\mathbb{P}(\mathbf{X} = \mathbf{x}) = \left(\frac{1}{\kappa}\right) \exp\left\{U(\mathbf{x})\right\}. \tag{8.4}$$

Here $U(\cdot)$ is called the *energy function* and

$$\kappa = \sum_{\mathbf{x}} \exp\left\{U(\mathbf{x})\right\}, \tag{8.5}$$

the *partition function*. Importantly, the energy function can be decomposed as a sum over cliques in G, in the form

$$U(\mathbf{x}) = \sum_{c \in \mathscr{C}} U_c(\mathbf{x}) \, , \tag{8.6}$$

where $\mathscr{C}$ denotes the set of all cliques of all sizes in G, and a clique of size 1 consists of just a single vertex $v \in V$.

In this abstract form, MRF models can involve extremely complicated expressions, a fact which, while arguably an indication of their richness on the one hand, can adversely impact both interpretability and computations. In practice, these models often are simplified by assumptions of homogeneity, in the sense that the form of the clique potentials U_c is assumed not to depend on the particular positions of the cliques $c \in \mathscr{C}$. Furthermore, usually cliques of only a limited size are defined to have non-zero partition functions U_c, which reduces the complexity of the decomposition in (8.6). This later step has direct implications on the nature of the assumed dependency in $\mathbf{X}$.

Here we will focus on a class of MRFs commonly used in network analysis for modeling binary vertex attribute data, like the indicators of protein function (e.g., ICSC) discussed previously. These models are sometimes referred to as *auto-logistic models*.[4]

8.3.2 Auto-Logistic Models

The class of auto-logistic models goes back to Besag [3], who suggested introducing the additional conditions on MRFs that (i) only cliques $c \in \mathscr{C}$ of size one or two have non-zero potential functions U_c, and (ii) the conditional probabilities in (8.3) have an exponential family form (i.e., a form like that in (6.1)). The first condition is sometimes referred to as 'pairwise-only dependence.' Under these conditions, the energy function takes the form

$$U(\mathbf{x}) = \sum_{i \in V} x_i H_i(x_i) + \sum_{\{i, j\} \in E} \beta_{ij} x_i x_j \, , \tag{8.7}$$

for some functions $H_i(\cdot)$ and coefficients $\{\beta_{ij}\}$. Besag called the class of Markov random field models with energy functions like that in (8.7) *auto-models*.

Now suppose that the X_i are binary random variables (i.e., taking on just the values zero and one). Under appropriate normalization conditions, the functions H_i

---

[4]For continuous-valued data, there is an analogous framework of so-called auto-Gaussian models, which bear close structural resemblance to the class of Gaussian graphical models we have seen in Sect. 7.3.3. We refer the reader to [13, Sect 8.3.1] for a brief discussion of these models, in the context of network analysis, and additional references.

can be made to only contribute to the expansion of $U(\mathbf{x})$ in (8.6) in a non-trivial fashion when $x_i = 1$, in which case (8.7) can be shown to be equivalent in form to

$$U(\mathbf{x}) = \sum_{i \in V} \alpha_i x_i + \sum_{\{i,j\} \in E} \beta_{ij} x_i x_j , \qquad (8.8)$$

for certain parameters $\{\alpha_i\}$. The resulting MRF model is called an *auto-logistic* model, because the conditional probabilities in (8.3) have the form

$$\mathbb{P}\left(X_i = 1 \mid \mathbf{X}_{\mathcal{N}_i} = \mathbf{x}_{\mathcal{N}_i}\right) = \frac{\exp\left(\alpha_i + \sum_{j \in \mathcal{N}_i} \beta_{ij} x_j\right)}{1 + \exp\left(\alpha_i + \sum_{j \in \mathcal{N}_i} \beta_{ij} x_j\right)} , \qquad (8.9)$$

indicating logistic regression of x_i on its neighboring x_j's.

Assumptions of homogeneity can further simplify this model. For example, specifying that $\alpha_i \equiv \alpha$ and $\beta_{ij} \equiv \beta$, the probability in (8.9) reduces to

$$\mathbb{P}_{\alpha,\beta}\left(X_i = 1 \mid \mathbf{X}_{\mathcal{N}_i} = \mathbf{x}_{\mathcal{N}_i}\right) = \frac{\exp\left(\alpha + \beta \sum_{j \in \mathcal{N}_i} x_j\right)}{1 + \exp\left(\alpha + \beta \sum_{j \in \mathcal{N}_i} x_j\right)} . \qquad (8.10)$$

This model can be read as dictating that the logarithm of the conditional odds that $X_i = 1$ scales linearly in the number of neighbors j of i with the value $X_j = 1$,

$$\log \frac{\mathbb{P}_{\alpha,\beta}\left(X_i = 1 \mid \mathbf{X}_{\mathcal{N}_i} = \mathbf{x}_{\mathcal{N}_i}\right)}{\mathbb{P}_{\alpha,\beta}\left(X_i = 0 \mid \mathbf{X}_{\mathcal{N}_i} = \mathbf{x}_{\mathcal{N}_i}\right)} = \alpha + \beta \sum_{j \in \mathcal{N}_i} x_j . \qquad (8.11)$$

Hence, we see that such (homogeneous) auto-logistic models effectively can be viewed as probabilistic extensions of nearest-neighbor methods. The R package **ngspatial** allows for the specification and fitting of such models.

```
#8.15 1 > library(ngspatial)
```

More precisely, it allows for models with both endogenous and exogenous effects, in which the logarithm of the conditional odds in (8.11) takes the form

$$\log \frac{\mathbb{P}_{\alpha,\beta}\left(X_i = 1 \mid \mathbf{X}_{\mathcal{N}_i} = \mathbf{x}_{\mathcal{N}_i}, \mathbf{Z}_i = \mathbf{z}_i\right)}{\mathbb{P}_{\alpha,\beta}\left(X_i = 0 \mid \mathbf{X}_{\mathcal{N}_i} = \mathbf{x}_{\mathcal{N}_i}, \mathbf{Z}_i = \mathbf{z}_i\right)} = \mathbf{z}_i^T \alpha + \beta \sum_{j \in \mathcal{N}_i} (x_j - \mu_j) . \qquad (8.12)$$

Here the $\mathbf{Z}_i$ are vectors of exogenous variables, indexed by vertex i, and α is a corresponding vector of coefficients. The values $\mu_j = \{1 + \exp(-\mathbf{z}_j^T \alpha)\}^{-1}$ are the expectations of the X_j under independence, i.e., when $\beta = 0$. The presence of the μ_j amounts to a centering of the model, which has been found to be useful for both interpretation (i.e., β in this model still reflects only dependence due to the underlying network, just as in (8.11)) and for improved fitting [5].

The specification of such models thus requires three pieces: the network process **X** to be modeled, the network G, and the set of relevant exogenous variables (if any). Returning to our problem of predicting the protein function ICSC in the network `ppi.CC`, the first two specifications are accomplished in **ngspatial** through the following assignments.

```
#8.16 1 > X <- V(ppi.CC.gc)$ICSC
      2 > A <- as_adjacency_matrix(ppi.CC.gc, sparse=FALSE)
```

The last specification depends on what additional information we wish to incorporate. For example, if, as in (8.11), we wish only to have an intercept, then we indicate this by

```
#8.17 1 > formula1 <- X~1
```

Alternatively, biology tells us that various types of protein-specific information besides interactions can be useful in predicting protein function. An example is information on the genetic sequence underlying that gene coding for a given protein. For instance, genetic motifs are short sequences of DNA thought to have biological significance, such as by influencing the spatial configuration of proteins. Indicators of the presence or absence of six such motifs are included with the network `ppi.CC` as vertex attribute variables, each starting with the letters 'IPR'. These are natural candidates for exogenous variables in our model.

```
#8.18 1 > gene.motifs <- cbind(V(ppi.CC.gc)$IPR000198,
      2 +                        V(ppi.CC.gc)$IPR000403,
      3 +                        V(ppi.CC.gc)$IPR001806,
      4 +                        V(ppi.CC.gc)$IPR001849,
      5 +                        V(ppi.CC.gc)$IPR002041,
      6 +                        V(ppi.CC.gc)$IPR003527)
      7 > formula2 <- X ~ gene.motifs
```

We will see below that they indeed have nontrivial predictive power.

8.3.3 Inference and Prediction for Auto-Logistic Models

As with our treatment of the nearest-neighbor method, we focus again on the task of prediction of network processes **X**. However, unlike previously, we require knowledge of the handful of parameters in our models in order to generate predictions. This will, of course, be true of Markov random field models in general. In the specific context of the auto-logistic models (8.12), it is the parameters α and β that are needed. Given measurements of **X**, we can try to infer these parameters from the data.

In principle, the task of inferring the vector (α^T, β) is most naturally approached through the method of maximum likelihood. But in practice this method often proves to be intractable. Consider, for example, the auto-logistic model without exogenous effects, in (8.10). The maximum likelihood estimate (MLE) of (α, β) is defined as the value $\left(\hat{\alpha}, \hat{\beta}\right)_{MLE}$ maximizing the log-likelihood $\log \mathbb{P}_{\alpha, \beta}(\mathbf{X} = \mathbf{x})$, a task that can be shown to be equivalent to maximizing the expression

$$\alpha M_1(\mathbf{x}) + \beta M_{11}(\mathbf{x}) - \kappa (\alpha, \beta) \ . \tag{8.13}$$

Here $M_1(\mathbf{x})$ is the number of vertices with the attribute value 1, $M_{11}(\mathbf{x})$ is twice the number of adjacent pairs of vertices where both have the attribute value 1, and $\kappa(\alpha, \beta)$ is the partition function for this model, corresponding to the generic function defined in (8.5). Unfortunately, calculation of $\kappa(\alpha, \beta)$ is prohibitive, as it requires evaluation of M_1 and M_{11} across all 2^{N_v} binary vectors $\mathbf{x}$ of length N_v, with respect to the network graph G.

The method of maximum pseudo-likelihood, originally proposed by Besag [4] for the analysis of spatial data, is a popular, computationally-feasible alternative to maximum likelihood in the context of MRF models. For the specific case of our auto-logistic models, instead of optimizing the *marginal* log-likelihood $\log \mathbb{P}_{\alpha, \beta}(\mathbf{X} = \mathbf{x})$, we instead seek to maximize the so-called *pseudo* log-likelihood

$$\sum_{i \in V} \log \mathbb{P}_{\alpha, \beta} \left(X_i = x_i \mid \mathbf{X}_{\mathcal{N}_i} = \mathbf{x}_{\mathcal{N}_i} \right) \ , \tag{8.14}$$

which is the logarithm of the product of the conditional probabilities of each observed x_i, given the values of its neighbors.

Importantly, these conditional probabilities do not involve the partition function $\kappa(\alpha, \beta)$. For the auto-logistic model without exogenous effects, in (8.10), the maximum pseudo-likelihood estimate (MPLE) can be shown to be that value $\left(\hat{\alpha}, \hat{\beta}\right)_{MPLE}$ which maximizes

$$\alpha M_1(\mathbf{x}) + \beta M_{11}(\mathbf{x}) - \sum_{i=1}^{N_v} \log \left[1 + \exp \left(\alpha + \beta \sum_{j \in \mathcal{N}_i} x_j \right) \right] \ . \tag{8.15}$$

While the solution to this optimization does not have a closed-form expression, the piece defined by the summation over vertices i, replacing $\kappa(\alpha, \beta)$ in (8.13), can now be computed easily. In fact, the overall estimate can be computed using standard software for logistic regression, with the N_v pairs $\left(x_i, \sum_{j \in \mathcal{N}_i} x_j \right)$ serving as response and predictor variables, respectively.

Maximum pseudo-likelihood estimates typically will differ from maximum likelihood estimates. Experience, however, has found them to be fairly accurate, as long as the dependencies inherent in the full joint distribution are not too substantial to be ignored. These estimates may be calculated for the centered auto-logistic model defined in (8.12) using the `autologistic` function in **ngspatial**.[5] For example, we can fit the simpler of our two models to the protein interaction data as follows.

```
#8.19  1  > m1.mrf <- autologistic(formula1, A=A,
       2  +    control=list(confint="none"))
```

With the resulting coefficients estimated[6] to be

```
#8.20  1  > m1.mrf$coefficients
       2  (Intercept)            eta
       3   0.2004949     1.1351942
```

we see, for example, that in these data the addition of one neighboring protein with the function ICSC is estimated to increase the log-odds of the ego protein having ICSC by a factor of roughly 1.135.

In order to obtain some sense as to how effective we might be in predicting ICSC using this model,[7] we consider the simple prediction rule that a protein have ICSC if the fitted probability $\mathbb{P}_{\hat{\alpha},\hat{\beta}}\left(X_i = 1 \mid \mathbf{X}_{\mathcal{N}_i} = \mathbf{x}_{\mathcal{N}_i}\right)$ is greater than 0.5,

```
#8.21  1  > mrf1.pred <- as.numeric((m1.mrf$fitted.values > 0.5))
```

which yields an error rate of roughly 20 %.

```
#8.22  1  > mean(as.numeric(mrf1.pred != V(ppi.CC.gc)$ICSC))
       2  [1] 0.2047244
```

This may be compared to the 25% error rate we witnessed with the nearest-neighbor method. However, with respect to the seven proteins that were discovered to have ICSC between 2007 and 2017, we find that this model and the nearest-neighbor method are similar, i.e, the same four new annotations are correctly predicted.

```
#8.23  1  > m1.mrf$fitted.values[V(ppi.CC.gc)$name %in% new.icsc]
       2  [1] 0.7519142 0.1658647 0.2184092 0.6451897
       3  [5] 0.9590030 0.2595863 0.3956048
```

---

[5]In fact, the function `autologistic` also implements a Bayesian approach to estimation of the parameters α and β. However, as this method is decidedly more computationally intensive, and its implementation requires the use of parallel processing, we discuss only the pseudo-likelihood method here, which is the default for `autologistic`.

[6]It is also possible to obtain approximate confidence intervals of α and β using `autologistic`. However, this option too requires the use of parallel processing and is thus omitted here to simplify our exposition.

[7]Note that if we observe only some of the elements of $\mathbf{X}$, say $\mathbf{X}^{obs} = \mathbf{x}^{obs}$, and we wish to predict the remaining elements $\mathbf{X}^{miss}$, then formally we should do so based on the distribution $\mathbb{P}_{\alpha,\beta}(\mathbf{X}^{miss}|\mathbf{X}^{obs} = \mathbf{x}^{obs})$. Explicit evaluation of this distribution will be prohibitive, but it is relatively straightforward to simulate from this distribution using the Gibbs sampler, a type of Markov chain Monte Carlo algorithm. See [13, Sect. 8.3.2.2] for a brief sketch of this approach.

The inclusion of gene motif information in our model

```
#8.24 1 > m2.mrf <- autologistic(formula2, A=A,
      2 +    control=list(confint="none"))
```

actually leads to an estimated network effect that is greater than in the previous model, i.e., approximately 1.30.

```
#8.25 1 > m2.mrf$coefficients
      2    (Intercept)    gene.motifs1    gene.motifs2    gene.motifs3
      3    5.081573e-02    1.876848e+00    1.875217e+01    1.875217e+01
      4    gene.motifs4    gene.motifs5    gene.motifs6             eta
      5    1.824990e+01    8.487244e-08   -1.837997e+01    1.297921e+00
```

And the error rate improves slightly.

```
#8.26 1 > mrf.pred2 <- as.numeric((m2.mrf$fitted.values > 0.5))
      2 > mean(as.numeric(mrf.pred2 != V(ppi.CC.gc)$ICSC))
      3 [1] 0.1889764
```

But perhaps most interestingly, this model appears to come much closer than the simpler model to correctly predicting ICSC function for five of the seven proteins of interest (using a threshold of 0.5).

```
#8.27 1 > m2.mrf$fitted.values[V(ppi.CC.gc)$name %in% new.icsc]
      2 [1] 0.7829254 0.4715219 0.4962188 0.6570828 0.7829254
      3 [6] 0.2175373 0.3510037
```

8.3.4   Goodness of Fit

Similar to as was discussed earlier in Chap. 6, in the context of modeling network graphs, it is important to assess the goodness-of-fit here too in the context of MRF models. Again, at this point in time, simulation appears to be the primary tool available for this purpose. Here, given a fitted MRF model, we would like to simulate realizations of the network process **X** from this model. Various statistics summarizing characteristics of these realizations may then be computed and compared to what results when the same is done for the original data.

The function `rautologistic` in the **ngspatial** package can be used to simulate realizations of centered autologistic models. The following code simulates 100 realizations from each of the two auto-logistic models considered above for the prediction of the protein function ICSC. For each realization **X**, the assortativity coefficient r_a is calculated with respect to the network `ppi.CC.gc`.

```
#8.28  1  > set.seed(42)          # random seed for rautologistic
       2  > ntrials <- 100
       3  > a1.mrf <- numeric(ntrials)
       4  > a2.mrf <- numeric(ntrials)
       5  > Z1 <- rep(1,length(X))
       6  > Z2 <- cbind(Z1, gene.motifs)
       7  > for(i in 1:ntrials){
       8  +    X1.mrf <- rautologistic(as.matrix(Z1), A=A,
       9  +                            theta=m1.mrf$coefficients)
      10  +    X2.mrf<- rautologistic(as.matrix(Z2), A=A,
      11  +                            theta=m2.mrf$coefficients)
      12  +    a1.mrf[i] <- assortativity(ppi.CC.gc, X1.mrf+1,
      13                                  directed=FALSE)
      14  +    a2.mrf[i] <- assortativity(ppi.CC.gc, X2.mrf+1,
      15                                  directed=FALSE)
      16  + }
```

The assortativity coefficient for the originally observed labels of ICSC function is roughly 0.37—fairly high, consistent with our findings that ICSC may be predicted reasonably well from protein–protein interactions.

```
#8.29  1  > assortativity(ppi.CC.gc, X+1, directed=FALSE)
       2  [1] 0.3739348
```

Comparing this value to the distribution of those values obtained under our two models, we find that it falls in the upper quartile.

```
#8.30  1  > summary(a1.mrf)
       2      Min. 1st Qu.  Median    Mean 3rd Qu.     Max.
       3  0.06736 0.22324 0.28733 0.27485 0.34203 0.46534
       4  > summary(a2.mrf)
       5      Min.  1st Qu.   Median    Mean  3rd Qu.     Max.
       6  -0.02656  0.20478  0.27621  0.26621  0.32519  0.45848
```

This suggests that the goodness-of-fit of our models, while not bad, can likely still be improved upon.

8.4 Kernel Methods

The probabilistic models we have just seen postulate a precise form for the dependency structure among vertex attributes X_i, with respect to the topology of the underlying graph G. But in some contexts, such as when prediction of unobserved vertex attributes is the only goal, it may be felt sufficient simply to 'learn' from the data a function relating the vertices to their attributes. The nearest-neighbor methods discussed in Sect. 8.2 in principle yield such a function, albeit implicitly. For a more explicit construction, a regression-based approach i.e., essentially regression on the graph G, would seem appealing. However, standard methods of regression, such as classical least squares regression, being set up as they are for relating response and predictor variables in Euclidean space, are not immediately applicable to graph-indexed data.

Kernel methods have been found to be useful for extending the classical regression paradigm to various settings with non-traditional data. At the most basic level, these methods consist of (i) a generalized notion of predictor variables (i.e., encoded in a so-called 'kernel'), and (ii) regression of a response on these generalized predictors using a penalized regression strategy. In this section we first introduce the notion of a kernel on a graph, and we then discuss the basic kernel approach to regression modeling in the context of graphs.

8.4.1 Designing Kernels on Graphs

At the heart of kernel methods is the notion of a kernel function. Broadly speaking, kernels can be thought of as functions that produce similarity matrices. The predictor variables used in the kernel regression are in turn derived from these similarity matrices. In the present context of kernel regression on network graphs, the kernel describes the similarity among vertices in the underlying graph G. Since G itself often is defined to represent such similarities, it is common to construct kernels that summarize the topology of G.

Formally, a function K, receiving vertex pairs (i, j) as input and returning a real value as output, is called a (*positive semi-definite*) *kernel* if, for each $m = 1, \ldots, N_v$ and subset of vertices $\{i_1, \ldots, i_m\} \in V$, the $m \times m$ matrix $\mathbf{K}^{(m)} = [K(i_j, i_{j'})]$ is symmetric and positive semi-definite.[8]

Although vertex proximity—and, hence, hopefully similarity—is naturally encoded in the adjacency matrix $\mathbf{A}$, it is more common to see the graph Laplacian used in the context of kernel regression. Recall from Sect. 4.4.2 that the graph Laplacian is defined as $\mathbf{L} = \mathbf{D} - \mathbf{A}$, where $\mathbf{D} = \mathsf{diag}\,[(d_v)]$. The *Laplacian kernel* is defined simply as the (pseudo)inverse of the Laplacian, i.e., $\mathbf{K} = \mathbf{L}^-$.

More precisely, write $\mathbf{L} = \Phi\Gamma\Phi^T$, where Φ and $\Gamma = \mathsf{diag}[(\gamma_i)]$ are $N_v \times N_v$ orthogonal and diagonal matrices, respectively, arising through the eigen-decomposition of $\mathbf{L}$. Then the (pseudo)inverse is given by

$$\mathbf{L}^- = \sum_{i=1}^{N_v} f(\gamma_i)\phi_i\phi_i^T , \qquad (8.16)$$

where ϕ_i is the i-th column of Φ and

$$f(\gamma) = \begin{cases} \gamma^{-1}, & \text{if } \gamma \neq 0, \\ 0, & \text{otherwise}. \end{cases} \qquad (8.17)$$

---

[8] A matrix $\mathbf{M} \in \mathbb{R}^{m \times m}$ is positive semi-definite if $\mathbf{x}^T\mathbf{M}\mathbf{x} \geq 0$, for all $\mathbf{x} \in \mathbb{R}^m$.

It is not difficult to show that $\mathbf{L}^-$ is symmetric and positive semi-definite, and hence a proper kernel matrix.

Why might $\mathbf{L}^-$ be a reasonable choice of kernel? We shall see shortly that in kernel regression the values of a vertex process $\mathbf{X}$ on a graph G are predicted through linear combinations of the eigen-vectors ϕ_i. That is, through quantities of the form $\mathbf{h} = \Phi\beta$. However, in order to better constrain the choice of such quantities, in fitting to data, penalized regression strategies are used, wherein $\beta^T \Gamma\beta = \sum_{i=1}^{N_v} \gamma_i \beta_i^2$ is encouraged to be 'small'. But

$$\beta^T \Gamma\beta = \beta^T \Phi^T \Phi \Gamma \Phi^T \Phi \beta$$
$$= \mathbf{h}^T \mathbf{L} \mathbf{h} \ . \tag{8.18}$$

Furthermore, it can be shown that

$$\mathbf{h}^T \mathbf{L} \mathbf{h} = \sum_{\{i,j\}\in E} (h_i - h_j)^2 \ . \tag{8.19}$$

So $\beta^T \Gamma\beta$ in (8.18) will be small if and only if $\mathbf{h}^T \mathbf{L} \mathbf{h}$ in (8.19) is small, which in turn is the case when the values of $\mathbf{h}$ assigned to vertices i and j adjacent in G are close, so as to reduce the magnitude of the differences $h_i - h_j$.

In other words, use of the kernel $\mathbf{K} = \mathbf{L}^-$ will encourage kernel regression to seek vectors $\mathbf{h} = \Phi\beta$ that are locally 'smooth' with respect to the topology of G. In this sense, therefore, kernel regression can be made to behave similarly on network graphs to the manner in which we have already seen nearest-neighbor and Markov random field methods to behave.

By way of illustration, consider again the network ppi.CC of protein interactions. In Fig. 8.3 are shown the weights $f(\gamma_i)$ defining the Laplacian kernel $\mathbf{K} = \mathbf{L}^-$ in (8.16) for the giant connected component of this network.

```
#8.31 1 > par(mfrow=c(1,1))
      2 > L <- as.matrix(laplacian_matrix(ppi.CC.gc))
      3 > e.L <- eigen(L)
      4 > nv <- vcount(ppi.CC.gc)
      5 > e.vals <- e.L$values[1:(nv-1)]
      6 > f.e.vals <- c((e.vals)^(-1), 0)
      7 > plot(f.e.vals, col="magenta", xlim=c(1, nv),
      8 +    xlab=c("Index i"), ylab=expression(f(gamma[i])))
```

Because this subgraph is connected, the smallest eigenvalue γ_{min} of $\mathbf{L}$ is zero, and all others are positive. Upon inversion of the nonzero eigenvalues, through application of the function f in (8.17), we see that there are a relatively small percentage (e.g., the last 20 or so) of rather large weights $f(\gamma_i)$, with the rest being comparatively much smaller.

As a result of this behavior of the weights, we know that the structure of $\mathbf{K}$ is largely governed by the structure of a correspondingly small percentage of eigen-vectors. In Fig. 8.4 is shown a visual representation of the eigen-vectors corresponding to the three largest weights. The first is produced through the following code.

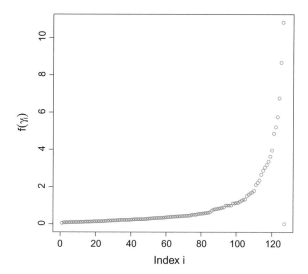

Fig. 8.3 Plot of the weights $f(\gamma_i)$ defining the Laplacian kernel $\mathbf{K} = \mathbf{L}^-$ in (8.16)

Fig. 8.4 Visual representation of the eigenvectors ϕ_i corresponding to the largest, second largest, and third largest (*left to right*) weights $f(\gamma_i)$ for the protein interaction network `ppi.CC.gc`. Negative values are shown in *blue*, and positive values, in *red*, with the area of each vertex proportional to the magnitude of its entry in the corresponding eigenvector

```
#8.32 1 > e.vec <- e.L$vectors[, (nv-1)]
      2 > v.colors <- character(nv)
      3 > v.colors[e.vec >= 0] <- "red"
      4 > v.colors[e.vec < 0] <- "blue"
      5 > v.size <- 15 * sqrt(abs(e.vec))
      6 > l <- layout_with_fr(ppi.CC.gc)
      7 > plot(ppi.CC.gc, layout=l, vertex.color=v.colors,
      8 +     vertex.size=v.size, vertex.label=NA)
```

The other two are produced similarly, with `nv-1`, in the first line, replaced by `nv-2` and `nv-3`, respectively. For each of these eigen-vectors we see evidence of large clusters of vertices with values of similar sign and magnitude, indicative of smooth behavior on the network graph.

We will use the R package **kernlab** to apply kernel regression methods to our problem of modeling and predicting processes on network graphs.

```
#8.33 1 > library(kernlab)
```

This package contains implementations of a variety of kernel-based machine learning methods. The Laplacian kernel for our protein–protein interaction network is constructed and declared a kernel object as follows.

```
#8.34 1 > K1.tmp <- e.L$vectors %*% diag(f.e.vals) %*%
      2 +    t(e.L$vectors)
      3 > K1 <- as.kernelMatrix(K1.tmp)
```

The notion of a kernel is quite general and so, not surprisingly, there are a variety of other ways that a kernel can be defined in our context. For example, Smola and Kondor [19] introduced a more general class of kernels, based on a certain notion of 'regularization,' wherein the function f in (8.17) is replaced by various other choices of smooth function. Alternatively, we may wish to combine multiple sources of information in constructing our kernel. A common approach to doing so is to encode each separate source of information into its own kernel and to then define the kernel **K** as a convex combination of those kernels (so as to maintain symmetry and positive definiteness).

For example, recall that inclusion of the gene motif information in the matrix `gene.motifs` was useful for predicting ICSC protein function with Markov random field models. A straightforward way in which to encode this information in a kernel is through the use of inner products, resulting in a so-called inner-product kernel.

```
#8.35 1 > K.motifs <- gene.motifs %*% t(gene.motifs)
```

Giving equal weight[9] to both this and our Laplacian kernel results in a kernel that now incorporates both endogenous and exogenous network information.

```
#8.36 1 > K2.tmp <- 0.5 * K1.tmp + 0.5 * K.motifs
      2 > K2 <- as.kernelMatrix(K2.tmp)
```

8.4.2   Kernel Regression on Graphs

We now formalize the notion of kernel regression on graphs and illustrate the type of performance that can be obtained in the context of our prediction problem. Let $G = (V, E)$ be a network graph and $\mathbf{X} = (X_1, \ldots, X_{N_v})$ a vertex attribute process. From the perspective of kernel regression, our goal is to learn from the data an appropriate function, say $\hat{h}$, mapping from V to $\mathbb{R}$, that describes well the manner in which attributes vary across the vertices. More precisely, given a kernel **K**, with

---

[9]More ambitiously, these weights in turn may be chosen in a data-adaptive fashion. See, for example, [7, 8, 14, 16, 20].

eigen-decomposition $\mathbf{K} = \Phi \Delta \Phi^T$, in kernel regression we seek to find an optimal choice of $\mathbf{h}$ within the class

$$\mathcal{H}_K = \left\{ \mathbf{h} \; : \; \mathbf{h} = \Phi \beta \text{ and } \beta^T \Delta^{-1} \beta < \infty \right\} , \qquad (8.20)$$

where $\mathbf{h}$ is an N_v-length vector.

In order to choose an appropriate element $\mathbf{h}$ in $\mathcal{H}_K$, say $\hat{\mathbf{h}}$, a penalized regression strategy is employed in kernel regression in an effort to enforce that $\hat{\mathbf{h}}$ both be close to the observed data and be sufficiently smooth (i.e., in the sense of (8.19) or, equivalently, (8.18), being small). Specifically, an estimate $\hat{\mathbf{h}} = \Phi \hat{\beta}$ is produced by selecting that $\hat{\beta}$ that minimizes

$$\sum_{i \in V^{obs}} C\left(x_i \; ; \; (\Phi \beta)_i\right) + \lambda \beta^T \Delta^{-1} \beta , \qquad (8.21)$$

where $V^{obs} \subseteq V$ denotes the set of vertices i at which we have observations $X_i = x_i$, $C(\cdot; \cdot)$ is a convex function that measures the loss incurred through predicting its first argument by its second, $(\Phi \beta)_i$ denotes that element of $\Phi \beta$ corresponding to $i \in V^{obs}$, and λ is a tuning parameter.

The optimization in (8.21) is a type of complexity-penalized estimation strategy. The role of the predictor variable is played by the columns of the matrix Φ (i.e., the eigenvectors of the kernel matrix $\mathbf{K}$), and that of the response variable, by the observed elements of $\mathbf{X}$. The loss captured by $C(\cdot; \cdot)$ encourages goodness of fit in the model, while the term $\beta^T \Delta^{-1} \beta$ penalizes excessive complexity, in the sense that eigenvectors with small eigenvalues are penalized more harshly than those with large eigenvalues. The parameter λ dictates the relative importance of the loss versus the complexity penalty, and is typically chosen in a data-dependent fashion.

For certain choices of loss, the optimal solution to the minimization of the expression in (8.21) has a closed form. In general, however, numerical methods must be used. The details of the implementation depend on the specific choices of loss function $C(\cdot; \cdot)$. In **kernlab**, for the problem of regression with binary response variables, a standard choice of loss function is implemented in the function ksvm. Specifically, the loss

$$C(x; h) = \left[\max\left(0, 1 - \tilde{x}h\right)\right]^2 \qquad (8.22)$$

is used, where $\tilde{x} = 2x - 1$, mapping the values $x = 0$ or 1 to -1 and $+1$, respectively. This choice corresponds to what is known as a 2-norm soft-margin support vector machine.

Returning to our motivating problem of predicting ICSC using our network of protein interactions, we use the Laplacian kernel K1 defined earlier as input to ksvm and extract the resulting fitted values, which ksvm produces in the form of 0 or 1 values.[10]

---

[10] Alternatively, using the option prob.model=TRUE, probabilities may be calculated.

```
#8.37  1 > m1.svm <- ksvm(K1, X, type="C-svc")
       2 > m1.svm.fitted <- fitted(m1.svm)
```

Comparing these fitted values to the originally observed indicators of ICSC function, we see that this kernel regression produces an error rate of 11%, roughly half of that of the analogous MRF model.

```
#8.38  1 > mean(as.numeric(m1.svm.fitted != V(ppi.CC.gc)$ICSC))
       2 [1] 0.1102362
```

Furthermore, the model accurately predicts ICSC function for four of the seven proteins of interest.

```
#8.39  1 > m1.svm.fitted[V(ppi.CC.gc)$name %in% new.icsc]
       2 [1] 1 1 1 0 1 0 0
```

If we now incorporate gene motif information as well, by using the kernel K2 instead of K1,

```
#8.40  1 > m2.svm <- ksvm(K2, X, type="C-svc")
```

we find that the overall error rate is again decreased by nearly half,

```
#8.41  1 > m2.svm.fitted <- fitted(m2.svm)
       2 > mean(as.numeric(m2.svm.fitted != V(ppi.CC.gc)$ICSC))
       3 [1] 0.06299213
```

although in this case, only two of the seven proteins of interest are predicted correctly.

```
#8.42  1 > m2.svm.fitted[V(ppi.CC.gc)$name %in% new.icsc]
       2 [1] 1 0 0 0 1 0 0
```

8.5   Modeling and Prediction for Dynamic Processes

As remarked earlier in this chapter, many processes defined on networks are more accurately thought of as dynamic, rather than static, processes. Examples include a cascade of failures (e.g., as an electrical power-grid strains under a heat-wave), the diffusion of knowledge (e.g., as a rumor spreads in a population), the search for information (e.g., as an Internet-based search engine formulates a response to a query), the spread of disease (e.g., as a virus propagates through a population of humans or computers), the synchronization of behavior (e.g., as neurons fire in the brain), and the interaction of 'species' in an environment (e.g., as genes in a cell auto-regulate among themselves).

Conceptually, we may think of processes like these as time-indexed vertex attribute processes $\mathbf{X}(t) = (X_i(t))_{i \in V}$, with t varying in a discrete or continuous manner over a range of times. Both deterministic and stochastic perspectives are commonly adopted for modeling such processes. Deterministic models are based on difference and differential equations, whereas stochastic models are based on time-indexed stochastic processes—usually Markov processes.[11]

While there has been a substantial amount of work done in the past 15–20 years on the mathematical and probabilistic modeling of dynamic processes on network graphs (see [2], for example, for a survey), there has been comparatively much less work on the statistics. As a result, our treatment of this topic here will be similarly limited in scope, with the goal being simply to take a quick peek at the modeling and simulation of such processes. We illustrate within the context of one particular class of dynamic process—epidemic processes.

8.5.1 Epidemic Processes: An Illustration

The term *epidemic* refers to a phenomenon that is prevalent in excess to what might be expected. It is most commonly used in the context of diseases and their dissemination throughout a population—such as with malaria, COVID-19, and AIDS—but it is also at times used more broadly in other contexts, such as in describing the spread of perceived problems in a society or the adoption of a commercial product.

Epidemic modeling has been an area of intense interest among researchers working on network-based dynamic process models. We begin our discussion here by briefly introducing a classical model from traditional (i.e., non-network) epidemic modeling. We then examine a network-based analogue.

Traditional Epidemic Modeling

The most commonly used class of continuous-time epidemic models is the class of *susceptible-infected-removed* (SIR) models. In this section, we will focus on the stochastic formulation of what is arguably the simplest member of this class—the so-called *general epidemic model*.

Imagine a (closed) population of, say, $N + 1$ elements such that, at any point in time t, there are some random number $N_S(t)$ of elements susceptible to infection (called 'susceptibles'), $N_I(t)$ elements infected (called 'infectives'), and $N_R(t)$ elements recovered and immune (or, alternatively, removed). Starting with one infective and N susceptibles, that is, with $N_I(0) = 1$ and $N_S(0) = N$, and letting s and i generically denote some numbers of susceptibles and infectives, respectively, it is assumed that the triple $(N_S(t), N_I(t), N_R(t))$ evolves according to the instantaneous transition probabilities

---

[11]Informally speaking, a *Markov process* is a stochastic process in which the future of the process at any given point in time depends only upon its present state and not its past.

$$\mathbb{P}\left(N_S(t+\delta t)=s-1,\,N_I(t+\delta t)=i+1\mid N_S(t)=s,\,N_I(t)=i\right)\approx\beta s i\delta t \qquad (8.23)$$

$$\mathbb{P}\left(N_S(t+\delta t)=s,\,N_I(t+\delta t)=i-1\mid N_S(t)=s,\,N_I(t)=i\right)\approx\gamma i\delta t$$

$$\mathbb{P}\left(N_S(t+\delta t)=s,\,N_I(t+\delta t)=i\mid N_S(t)=s,\,N_I(t)=i\right)\approx1-(\beta s+\gamma)i\delta t\,,$$

where δt refers to the usual infinitesimal and the role of $N_R(t)$ is omitted due to the constraint $N_S(t)+N_I(t)+N_R(t)=N+1$.

The above model states that, at any given time t, a new infective will emerge from among the susceptibles (due to contact with and infection by one of the infectives) with instantaneous probability proportional to the product of the number of suscepti- bles s and the number of infectives i. Similarly, infectives recover with instantaneous probability proportional to i. These probabilities are scaled by the parameters β and γ, usually referred to as the infection and recovery rates, respectively. The product form for the probability with which infectives emerge corresponds to an assumption of 'homogeneous mixing' (or 'mass-action,' in chemistry) among members of the population, which asserts that the population is (i) homogeneous and (ii) well mixed, in the sense that any pair of members are equally likely to interact with each other.

An equivalent formulation of the model states that, given s susceptibles and i infectives at time t, the process remains in the state (s,i) for an amount of time distributed as an exponential random variable, with rate $(\beta s+\gamma)i$. A transition then occurs, which will be either to the state $(s-1,i+1)$, with probability $\beta s/[(\beta s+\gamma)i]$, or to the state $(s,i-1)$, with probability $\gamma i/[(\beta s+\gamma)i]$.

Figure 8.5 shows a schematic characterization of the typical behavior of our stochastic SIR process under simulation.[12] Starting with a population composed almost entirely of susceptibles and just a few infectives, we see an initial exponential increase in the number of infectives and a corresponding decrease in the number of susceptibles. This initial period is followed by a peak in the number of infec- tives, after which this number decays exponentially, as the supply of susceptibles is depleted and the infectives recover.

Unfortunately, despite the fact that the general epidemic SIR model captures the gross characteristics of a canonical epidemic, the underlying assumption of homo- geneous mixing is admittedly simple and, for many diseases, too poor of an approx- imation to reality. A key element lacking from these models is the natural structure often inherent to populations. Such structure might derive from spatial proximity (e.g., diseases of plants, in which infection occurs through the help of carriers over short distances), social contact (e.g., sexual contact in the transmission of AIDS), or demographics (e.g., households, age brackets, etc.). More sophisticated models assume contact patterns that take into account such structure(s) within the population of interest. And frequently it is convenient to represent these patterns in the form of a graph.

---

[12]Tools in the R package **EpiModel** may be used to generate and analyze such curves, as well as curves corresponding to individual epidemic processes. As our primary focus is on network-based epidemics, however, we do not explore those tools here.

Fig. 8.5 Schematic characterization of an SIR process, showing how the relative proportions of those susceptible (*green*), infective (*red*), and removed (*yellow*) vary over time

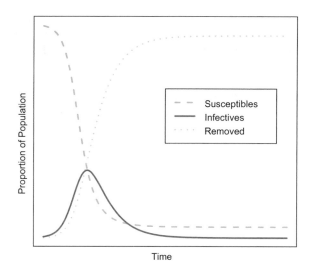

Network-Based Epidemic Modeling

Let G be a network graph describing the contact structure among N_v elements in a population. We assume that initially, at time $t = 0$, one vertex is infected and the rest are susceptible. Infected vertices remain infected for an amount of time distributed exponentially, with rate γ, after which they are considered recovered. During the infectious period a vertex has infectious contacts independently[13] with each neighbor, according to an exponential distribution with rate β, where an infectious contact automatically results in infection if the other individual is susceptible. Define $X_i(t) = 0, 1,$ or 2, according to whether vertex i is susceptible, infected, or removed at time t, respectively.

Let $\mathbf{X}(t) = (X_i(t))_{i \in V}$ be the resulting (continuous) time-indexed process[14] on the network graph G. Denote by $\mathbf{x}$ the state of the process at a given time t (i..e, the particular pattern of 0's, 1's, and 2's across the vertices in G at time t). Successive changes of states, say from $\mathbf{x}$ to $\mathbf{x}'$, will involve a change in one and only one element at a time. Suppose that $\mathbf{x}$ and $\mathbf{x}'$ differ in the i-th element. Then it may be shown that the model just described is equivalent to specifying that

$$\mathbb{P}\left(\mathbf{X}(t+\delta t) = \mathbf{x}' \mid \mathbf{X}(t) = \mathbf{x}\right) \approx \begin{cases} \beta M_i(\mathbf{x})\delta t , & \text{if } x_i = 0 \text{ and } x_i' = 1 , \\ \gamma \delta t , & \text{if } x_i = 1 \text{ and } x_i' = 2 , \\ 1 - \left[\beta M_i(\mathbf{x}) + \gamma\right]\delta t , & \text{if } x_i = 2 \text{ and } x_i' = 2 , \end{cases}$$
$$(8.24)$$

---

[13]Technically, independence pertains only to sufficiently small time intervals. Over the life of the epidemic, the events that an element in the population infects two neighbors are not independent, but rather positively correlated.

[14]Formally, $\mathbf{X}(t)$ constitutes a continuous-time Markov chain.

where we define $M_i(\mathbf{x})$ to be the number of neighbors $j \in \mathcal{N}_i$ for which $x_j = 1$ (i.e., the number of neighbors of i infected at time t). Our network-based analogue of the traditional SIR process then follows by defining the processes $N_S(t)$, $N_I(t)$, and $N_R(t)$, counting the numbers of susceptible, infective, and removed vertices at time t, respectively, in analogy to the traditional case.

In light of the expressions in (8.24), we can expect the characteristics of the processes $N_S(t)$, $N_I(t)$, and $N_R(t)$ to be affected to at least some extent by the characteristics of the network graph G. Simulation can be used to confirm this expectation.

We begin by generating examples of three different random graphs introduced earlier, in Chap. 5.

```
#8.43  1 > set.seed(42)
       2 > gl <- list()
       3 > gl$ba <- sample_pa(250, m=5, directed=FALSE)
       4 > gl$er <- sample_gnm(250, 1250)
       5 > gl$ws <- sample_smallworld(1, 250, 5, 0.01)
```

The parameters have been chosen so as to guarantee graphs of the same number of vertices and roughly the same average degree (i.e., around 10), since both are basic characteristics expected to fundamentally affect the progression of an epidemic in ways that do not reflect interesting differences in topology.

Setting the infection rate to be $\beta = 0.5$, and the recovery rate, to be $\gamma = 1.0$,

```
#8.44  1 > beta <- 0.5
       2 > gamma <- 1
```

we use the function `sir` in **igraph** to produce

```
#8.45  1 > ntrials <- 100
```

simulated epidemics on each network.

```
#8.46  1 > sim <- lapply(gl, sir, beta=beta, gamma=gamma,
       2 +    no.sim=ntrials)
```

The output from each simulation is an *sir* object, containing information about the times at which changes of states occurred and the values of the processes $N_S(t)$, $N_I(t)$, and $N_R(t)$ at those times.

The results of plotting the total number of infectives $N_I(t)$ for each of these three networks is shown in the first three panels of Fig. 8.6.

```
#8.47  1 > plot(sim$er)
       2 > plot(sim$ba, color="palegoldenrod",
       3 +    median_color="gold", quantile_color="gold")
       4 > plot(sim$ws, color="pink", median_color="red",
       5 +    quantile_color="red")
```

Individual simulation paths, as well as their medians and 10 and 90% quantiles, are shown in each panel.

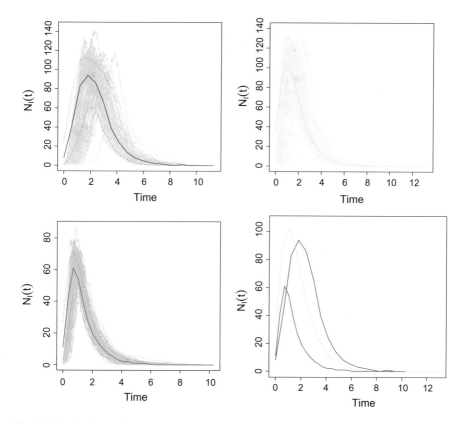

Fig. 8.6 Realizations of the number of infectives $N_I(t)$ for the network-based SIR process simulated on an Erdös-Rényi random graph (*blue*), a Barabási-Albert random graph (*yellow*), and a Watts-Strogatz 'small-world' random graph (*red*). *Darker curves* indicate the median (*solid*) and the 10th and 90th percentile (*dotted*), over a total of 100 epidemics (*shown in light curves*). The three median functions are compared in the lower right-hand plot

For all three network topologies we observe an exponential rise and decay that is qualitatively similar to that of a traditional SIR epidemic process, as shown in Fig. 8.5. Nevertheless, it is clear that they also differ in important ways, driven by differences in network topology. For example, the epidemic peaks earlier in the small-world network and the Barabási–Albert network than in the Erdős and Rényi network, yet the latter two networks yield substantially more infected nodes than the former. These differences may be better seen plotting the median of the curves $N_I(t)$, for each graph, on one plot.

```
#8.48  1  > x.max <- max(sapply(sapply(sim, time_bins), max))
       2  > y.max <- 1.05 * max(sapply(sapply(sim, function(x)
       3  +   median(x)[["NI"]]), max, na.rm=TRUE))
       4
       5  > plot(time_bins(sim$er), median(sim$er)[["NI"]],
       6  +   type="l", lwd=2, col="blue", xlim=c(0, x.max),
       7  +   ylim=c(0, y.max), xlab="Time",
       8  +   ylab=expression(N[I](t)))
       9  > lines(time_bins(sim$ba), median(sim$ba)[["NI"]],
      10  +   lwd=2, col="gold")
      11  > lines(time_bins(sim$ws), median(sim$ws)[["NI"]],
      12  +   lwd=2, col="red")
      13  > legend("topright", c("ER", "BA", "WS"),
      14  +   col=c("blue", "gold", "red"), lty=1)
```

8.6   Additional Reading

General discussion of nearest neighbor methods may be found in Hastie et al. [10, Sect. 2.3.2]. For an extensive development of Markov random fields, see Cressie [6, Sect. 6.4], where the emphasis is on modeling data on spatial lattices. Standard references on kernel methods include Schölkopf and Smola [17] and Shawe-Taylor and Cristianini [18]. Finally, for a general introduction to epidemic modeling, see Daley and Gani [9], for example, or Anderson and May [1], for a more in-depth treatment.

References

1. R. Anderson and R. May, *Infectious Diseases of Humans: Dynamics and Control*. Oxford: Oxford University Press, 1991.
2. A. Barrat, M. Barthélemy, and A. Vespignani, *Dynamical Processes on Complex Networks*. New York: Cambridge University Press, 2008.
3. J. Besag, "Spatial interaction and the statistical analysis of lattice systems," *Journal of the Royal Statistical Society, Series B*, vol. 36, no. 2, pp. 192–236, 1974.
4. J. Besag, "Statistical analysis of non-lattice data," *The Statistician*, vol. 24, no. 3, pp. 179–195, 1975.
5. P. C. Caragea and M. S. Kaiser, "Autologistic models with interpretable parameters," *Journal of Agricultural, Biological, and Environmental Statistics*, vol. 14, no. 3, pp. 281–300, 2009.
6. N. Cressie, *Statistics for Spatial Data*. New York: Wiley-Interscience, 1993.
7. N. Cristianini, J. Kandola, A. Elisseeff, and J. Shawe-Taylor, "On kernel-target alignment," in *Innovations in Machine Learning: Theory and Applications*, D. Holmes and L. Jain, Eds. New York: Springer-Verlag, 2006.
8. N. Cristianini, J. Shawe-Taylor, A. Elisseeff, and J. Kandola, "On kernel-target alignment," *Advances in Neural Information Processing Systems 14*, 2002.
9. D. Daley and J. Gani, *Epidemic Modeling*. New York: Cambridge University Press, 1999.
10. T. Hastie, R. Tibshirani, and J. Friedman, *The Elements of Statistical Learning*. New York: Springer, 2001.
11. E. Ising, "Beitrag zur theorie des ferromagnetismus," *Zeit. fur Physik*, vol. 31, pp. 253–258, 1925.

12. X. Jiang, N. Nariai, M. Steffen, S. Kasif, and E. Kolaczyk, "Integration of relational and hierarchical network information for protein function prediction," *BMC Bioinformatics*, vol. 9, no. 350, 2008.

13. E. Kolaczyk, *Statistical Analysis of Network Data: Methods and Models*. Springer Verlag, 2009.

14. G. Lanckriet, N. Cristianini, P. Bartlett, L. El Ghaoui, and M. Jordan, "Learning the kernel matrix with semidefinite programming," *Journal of Machine Learning Research*, vol. 5, pp. 27–72, 2004.

15. S. Li, *Markov Random Field Modeling in Computer Vision*. New York: Springer-Verlag, 1995.

16. Y. Lin and H. Zhang, "Component selection and smoothing in multivariate nonparametric regression," *Annals of Statistics*, vol. 34, no. 5, pp. 2272–2297, 2006.

17. B. Schölkopf and A. Smola, *Learning with Kernels: Support Vector Machines, Regularization, Optimization, and Beyond*. Cambridge, MA: MIT Press, 2002.

18. J. Shawe-Taylor and N. Cristianini, *Kernel Methods for Pattern Analysis*. Cambridge University Press, 2004.

19. A. Smola and R. Kondor, "Kernels and regularization on graphs," in *Proceedings of the 16th Annual Conference on Learning Theory (COLT)*, 2003.

20. H. Zhang and Y. Lin, "Component selection and smoothing for nonparametric regression in exponential families," *Statistica Sinica*, vol. 16, pp. 1021–1042, 2006.

Chapter 9
Analysis of Network Flow Data

9.1 Introduction

Many networks serve as conduits—either literally or figuratively—for *flows*, in the sense that they facilitate the movement of something, such as materials, people, or information. For example, transportation networks (e.g., of highways, railways, and airlines) support flows of commodities and people, communication networks allow for the flow of data, and networks of trade relations among nations reflect the flow of capital. We will generically refer to that of which a flow consists as *traffic*.

Flows are at the heart of the form and function of many networks, and understanding their behavior is often a goal of primary interest. Much of the quantitative work in the literature on flows is concerned with various types of problems involving, for example, questions of network design, provisioning, and routing, and their solutions have involved primarily tools in optimization and algorithms. Questions that are of a more statistical nature typically involve the modeling and prediction of network flow volumes, based on relevant data, and it is upon these topics that we will focus here.

Let $G = (V, E)$ be a network graph. Since flows have direction, from an origin to a destination, formally G will be a digraph. Following the convention in this literature, we will refer to the (directed) edges in this graph as links. Traffic typically passes over multiple links in moving between origin and destination vertices. A quantity of fundamental interest in the study of network flows is the so-called *origin-destination* (OD) matrix, which we will denote by $\mathbf{Z} = [Z_{ij}]$, where Z_{ij} is the total volume of traffic flowing from an origin vertex i to a destination vertex j in a given period of time. The matrix $\mathbf{Z}$ is also sometimes referred to as the *traffic matrix*.

In this chapter we will examine two statistical problems that arise in the context of network flows, distinguished by whether the traffic matrix $\mathbf{Z}$ is observed or to be predicted. More specifically, in Sect. 9.2, we consider the case where it is possible to observe the entire traffic matrix $\mathbf{Z}$ and it is of interest to model these observations. Modeling might be of interest both to obtain an understanding as to how potentially relevant factors (e.g., costs) affect flow volumes and to be able to make predictions

© Springer Nature Switzerland AG 2020   169
E. D. Kolaczyk and G. Csárdi, *Statistical Analysis of Network Data with R*, Use R!,
https://doi.org/10.1007/978-3-030-44129-6_9

of future flow volumes. A class of models commonly used for such purposes are the so-called gravity models. In Sect. 9.3, we consider the problem of traffic matrix estimation. Here the case is assumed to be such that it is difficult or impossible to observe the traffic matrix entries Z_{ij} directly. Nevertheless, given sufficient information on marginal flow volumes (i.e., in the form of total volumes through vertices or over links) and on the routing of flows between origins and destinations, it is often possible to generate accurate predictions of $\mathbf{Z}$ using statistical methods.

9.2   Modeling Network Flows: Gravity Models

Gravity models are a class of models, developed largely in the social sciences, for describing aggregate levels of interaction among the people of different populations. They have traditionally been used most in areas like geography, economics, and sociology, for example, but also have found application in other areas of the sciences, such as hydrology and the analysis of computer network traffic. Our interest in gravity models in this chapter will be for their use in contexts where the relevant traffic flows are over a network of sorts.

The term 'gravity model' derives from the fact that, in analogy to Newton's law of universal gravitation, it is assumed that the interaction among two populations varies in direct proportion to their size, and inversely, with some measure of their separation. The concept goes back at least to the work of Carey [2] in the 1850's, but arguably was formulated in the strictest sense of the analogy by Stewart [9] in 1941, and has since been developed substantially in the past 50 years. Here we focus on a certain general version of the gravity model and corresponding methods of inference.

9.2.1   Model Specification

We let $\mathscr{I}$ and $\mathscr{J}$ represent sets of origins i and destinations j in V, of cardinality $I = |\mathscr{I}|$ and $J = |\mathscr{J}|$, respectively. Furthermore, Z_{ij} denotes—as defined in the introduction—a measure of the traffic flowing from $i \in \mathscr{I}$ to $j \in \mathscr{J}$ over a given period of time. Suppose that we are able to observe $\mathbf{Z} = [Z_{ij}]$ in full.

For example, the data set `calldata` contains a set of data for phone traffic between 32 telecommunication districts in Austria throughout a period during the year 1991.

```
#9.1 1 > library(sand)
     2 > data(calldata)
     3 > names(calldata)
     4 [1] "Orig"     "Dest"     "DistEuc" "DistRd"   "O.GRP"
     5 [6] "D.GRP"    "Flow"
```

These data, first described[1] by Fischer and Gopal [5], consist of $32 \times 31 = 992$ flow measurements z_{ij}, $i \neq j = 1, \ldots, 32$, capturing contact intensity over the period studied. In addition to basic labels of origin and destination region for each flow, there is information on the gross regional product (GRP) for each region, which may serve as a proxy for economic activity and income, both of which are relevant to business and private phone calls. Finally, there are two sets of measurements reflecting notions of distance between regions, one based on Euclidean distance and the other road-based.

As the original data in the variable `Flow` are in units of erlang (i.e., number of phone calls, including faxes, times the average length of the call divided by the duration of the measurement period), and the gravity modeling frameworks we will describe assume units of counts, we convert these data to quasi-counts, for the purposes of illustration.

```
#9.2 1 > min.call <- min(calldata$Flow)
     2 > calldata$FlowCnt <- round(5 * calldata$Flow / min.call)
```

These flow counts may be thought of as corresponding to (directed) edge weights in a graph, where the vertices represent telecommunication districts.

```
#9.3 1 > W <- xtabs(FlowCnt ~ Orig + Dest, calldata)
     2 > g.cd <- graph_from_adjacency_matrix(W, weighted=TRUE)
```

A plot of the resulting network graph is shown in Fig. 9.1.

```
#9.4 1  > in.flow <- strength(g.cd, mode="in")
     2  > out.flow <- strength(g.cd, mode="out")
     3  > vsize <- sqrt(in.flow + out.flow) / 100
     4  > pie.vals <- lapply((1:vcount(g.cd)),
     5  +    function(i) c(in.flow[i], out.flow[i]))
     6  > ewidth <- E(g.cd)$weight / 10^5
     7  > set.seed(42)
     8  > plot(g.cd, vertex.size=vsize, vertex.shape="pie",
     9  +    vertex.pie=pie.vals, edge.width=ewidth,
     10 +    edge.arrow.size=0.1)
```

Gravity models have been found to be useful in modeling such data. The *general gravity model* specifies that the traffic flows Z_{ij} be in the form of counts, with independent Poisson distributions and mean functions of the form

$$\mathbb{E}(Z_{ij}) = h_O(i) \, h_D(j) \, h_S(\mathbf{c}_{ij}) \,, \tag{9.1}$$

where h_O, h_D, and h_S are positive functions, respectively, of the origin i, the destination j, and a vector $\mathbf{c}_{ij}$ of K so-called separation attributes. On a logarithmic scale, we therefore have

$$\log \mathbb{E}(Z_{ij}) = \log h_O(i) + \log h_D(j) + \log h_S(\mathbf{c}_{ij}) \,. \tag{9.2}$$

---

[1]Data were originally collected by Manfred Fischer and Petra Staufer.

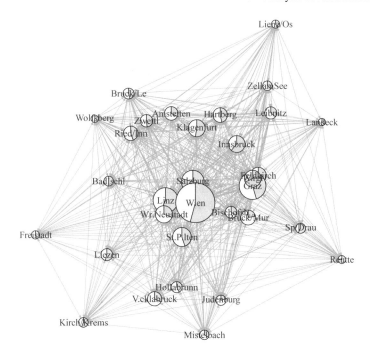

Fig. 9.1 Network representation of Austrian call network data flows in `calldata`. Vertex area reflects total flow volume in or out of a telecommunication region, with relative proportion of in- and out-flows indicated in *white* and *blue*, respectively. Width of links reflects volume of flow from origin to destination

As we shall see momentarily, common specifications of the functions h_O, h_D, and h_S tend to lead to expressions on the right-hand side of (9.2) that are linear in some unknown parameters. This log-linear form in turn facilitates the use of log-linear methods for statistical inference on the model parameters, as we discuss below in Sect. 9.2.2.

In general, the K elements of $\mathbf{c}_{ij}$ are chosen to quantify some notion(s) of separation ascribed to the origin-destination pair (i, j), often based on concepts of 'distance' or 'cost' of either a literal or figurative nature. The functions h_O and h_D are sometimes referred to as the *origin* and *destination functions*, respectively, while h_S is commonly called the *separation* or *deterrence function*. Often the separation function is constrained to be non-increasing in the elements of $\mathbf{c}_{ij}$.

An early and now classical example of the gravity model is that of Stewart [9], proposed in connection with his theory of 'demographic gravitation,' which specifies that

$$\mathbb{E}(Z_{ij}) = \gamma \ \pi_{O,i} \ \pi_{D,j} \ d_{ij}^{-2} \ , \tag{9.3}$$

where $\pi_{O,i}$ and $\pi_{D,j}$ are measures of the origin and destination population sizes, respectively, for two geographical regions i and j, and d_{ij} is a measure of distance between the centers of these regions. This formulation is completely analogous to

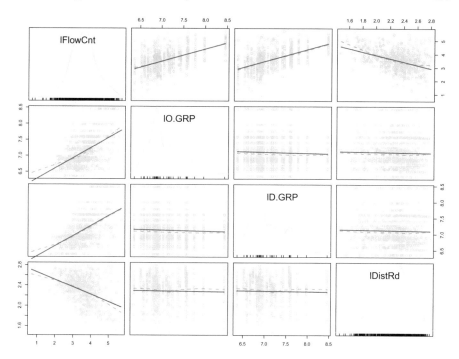

Fig. 9.2 Austrian call data. Scatterplots are shown for call flow volume versus each of origin GRP, destination GRP, and distance, along the top row, and for the latter three variables against each other, in the other rows. All axes are on log-log scales. Superimposed on each scatterplot are two lines, for descriptive purposes, showing fits based on simple linear regression (*red*) and a nonparametric smoother (*gold*). Density plots for each of the four variables are shown along the diagonal

Newton's universal law, right down to the use of a 'demographic gravitational constant' γ. However, unlike Newton's law, neither empirical evidence nor theoretical arguments suggest that this form of the gravity model is strictly accurate in practical contexts, where somewhat more flexible forms are used.

For example, consider again the Austrian call data. The gross regional products O.GRP and D.GRP can be used in analogy to population size, and we use the road-based measure of distance DistRd. A comparison of these variables against flow volume, on a logarithmic scale, is shown in Fig. 9.2.

```
#9.5  1  > calldata$lFlowCnt <- log(calldata$FlowCnt, 10)
      2  > calldata$lO.GRP <- log(calldata$O.GRP, 10)
      3  > calldata$lD.GRP <- log(calldata$D.GRP, 10)
      4  > calldata$lDistRd <- log(calldata$DistRd, 10)
      5  >
      6  > library(car)
      7  > scatterplotMatrix( ~ lFlowCnt + lO.GRP + lD.GRP +
      8  +    lDistRd, data=calldata, regLine=list(col="red"),
      9  +    smooth=list(spread=FALSE,col.smooth="goldenrod"),
     10  +    col="powderblue")
```

It is evident from examination of the scatterplots in this figure that there is a reasonably strong relationship between call volume and the origin GRP, destination GRP, and distance. Moreover, this relationship is fairly linear (i.e., the solid red lines, produced by ordinary least-squares, lie close to the dotted gold lines, which are the result of a nonparametric smoother) and is increasing in origin and destination GRP and decreasing in distance. These observations suggest that a model of the form

$$\mathbb{E}(Z_{ij}) = \gamma \left(\pi_{O,i}\right)^{\alpha} \left(\pi_{D,j}\right)^{\beta} \left(c_{ij}\right)^{\theta} \tag{9.4}$$

might be reasonable, where $\pi_{O,i}$ is the GRP of origin i, $\pi_{D,j}$ is the GRP of destination j, and c_{ij} is the distance from origin i to destination j. That is, a simple loglinear model of the form

```
#9.6 1 > formula.s <- FlowCnt ~ lO.GRP + lD.GRP + lDistRd
```

seems sensible. This choice corresponds to origin and destination functions of the form

$$h_O(i) = \left(\pi_{O,i}\right)^{\alpha} \quad \text{and} \quad h_D(j) = \left(\pi_{D,j}\right)^{\beta} \ , \tag{9.5}$$

and separation function,

$$h_S(c_{ij}) = \left(c_{ij}\right)^{\theta} \tag{9.6}$$

in the general gravity model.

Alternatively, the values $\{h_O(i)\}_{i \in \mathscr{I}}$ and $\{h_D(j)\}_{j \in \mathscr{J}}$ are often simply treated as a collection of $I + J$ unknown parameters. For instance, adopting this specification for the Austrian call data, while maintaining the same choice of separation function, yields the following more general formula.

```
#9.7 1 > formula.g <- FlowCnt ~ Orig + Dest + lDistRd
```

For a comprehensive treatment of additional variations on the general gravity model, including versions that relax the assumption of independence between origin and destination effects implicit in the product form $h_O(i) \times h_D(j)$, see Sen and Smith [8].

9.2.2 Inference for Gravity Models

In light of the specification that the Z_{ij} be independent Poisson random variables with means $\mu_{ij} = \mathbb{E}(Z_{ij})$, statistical inference in the general gravity model is most naturally approached through likelihood-based methods. Moreover, as remarked already, typical specifications of the general gravity model are in fact log-linear models, which in turn are a specific instance of the class of generalized linear models. See, for example, McCullagh and Nelder [7, Sect. 6]. As a result, these models often can be fit using standard software.

To be concrete, consider the model underlying `formula.g` above. This model can be expressed in the form

$$\log \mu_{ij} = \alpha_i + \beta_j + \theta^T \mathbf{c}_{ij} ,\tag{9.7}$$

where $\alpha_i = \log h_O(i)$, $\beta_j = \log h_D(j)$, and θ, $\mathbf{c}_{ij}$ are vectors of length K (with $K = 1$ in the context of our call data example). Other cases, such as when the origin and destination functions are parameterized, as in the expression (9.4) underlying `formula.s` above, may be handled similarly.

Let $\mathbf{Z} = \mathbf{z}$ be an $(IJ) \times 1$ vector of observations of the flows Z_{ij}, which for convenience are usually ordered by origin i, and by destination j within origin i (as is the case with the data set `calldata`). The relevant portion of the Poisson log-likelihood for μ takes the form

$$\ell(\mu) = \sum_{i,j \in \mathcal{I} \times \mathcal{J}} z_{ij} \log \mu_{ij} - \mu_{ij} .\tag{9.8}$$

Substituting the gravity model (9.7), taking partial derivatives with respect to the parameters α_i, β_j, and θ_k, setting the resulting equations equal to zero, and simplifying, it can be shown that the maximum likelihood estimates for these parameters must yield estimates $\hat{\mu}_{ij} = \hat{\alpha}_i \hat{\beta}_j \exp(\hat{\theta}^T \mathbf{c}_{ij})$, for μ_{ij} satisfying the equations

$$\hat{\mu}_{i+} = z_{i+}, \text{ for } i \in \mathcal{I} \quad \text{and} \quad \hat{\mu}_{+j} = z_{+j}, \text{ for } j \in \mathcal{J}\tag{9.9}$$

$$\sum_{i,j \in \mathcal{I} \times \mathcal{J}} c_{ij;k} \hat{\mu}_{ij} = \sum_{i,j \in \mathcal{I} \times \mathcal{J}} c_{ij;k} z_{ij}, \text{ for } k = 1, \ldots, K ,\tag{9.10}$$

where $\hat{\mu}_{i+} = \sum_{j \in \mathcal{J}} \mu_{ij}$ and $\hat{\mu}_{+j} = \sum_{i \in \mathcal{I}} \mu_{ij}$, and the z_{i+} and z_{+j} are defined similarly. Here $c_{ij;k}$ denotes the kth element of $\mathbf{c}_{ij}$.

Under mild conditions, these estimates will be well defined, and furthermore, the values $\hat{\theta}_k$ and $\hat{\mu}_{ij}$ will be unique. The values $\hat{\alpha}_i$ and $\hat{\beta}_j$ will be unique only up to a constant, due to the fact that the underlying model is over-parameterized by one degree of freedom.[2] Note that we are assuming here, without loss of generality, that origins i and destinations j for which $z_{i+} = 0$ or $z_{+j} = 0$ are dropped, as they contribute nothing to the analysis.

Various algorithms may be used to calculate the maximum likelihood estimates. Most straightforward is to use standard software for fitting log-linear models, usually available as an option in routines for fitting generalized linear models. These

---

[2]More precisely, we can write (9.7) in the form $\log(\mu) = \mathbf{M}\gamma$, where $\mathbf{M}$ is an $(IJ) \times (I + J + K)$ matrix, and $\gamma = (\alpha_1, \ldots, \alpha_I, \beta_1, \ldots, \beta_J, \theta_1, \ldots, \theta_K)^T$ is an $(I + J + K) \times 1$ vector. The first $I + J$ columns of $\mathbf{M}$ are binary vectors, indicating the appropriate origin and destination for each entry of μ, and are redundant in that both the first I and the next J sum to the unit vector. The last K columns correspond to the K variables defining the $\mathbf{c}_{ij}$. Assuming that the latter are linearly independent of themselves and of the former, the rank of $\mathbf{M}$ will be $(I + J - 1) + K$. See Sen and Smith [8, Sect. 5.2].

procedures are based on an iteratively re-weighted least-squares algorithm, like that used in logistic regression, which derives from application of the Newton-Raphson algorithm. See, for example, McCullagh and Nelder [7, Sect. 2.5] for details. Standard output includes parameter estimates and approximate standard errors, where the latter are driven by the usual arguments for asymptotic normality of maximum likelihood estimators.

To illustrate, we use the function `glm` in the R base package to fit our simple (i.e,. `formula.s`) and more general (i.e., `formula.g`) specifications of gravity models for the Austrian call data.

```
#9.8 1 > gm.s <- glm(formula.s, family="poisson", data=calldata)
     2 > gm.g <- glm(formula.g, family="poisson", data=calldata)
```

The results of the fitted model `gm.s` are as follows.

```
#9.9  1 > summary(gm.s)
      2
      3 Call:
      4 glm(formula = formula.s, family = "poisson", data = calldata)
      5
      6 Deviance Residuals:
      7      Min        1Q    Median        3Q       Max
      8 -475.06    -54.16    -29.20     -2.09   1149.93
      9
     10 Coefficients:
     11                  Estimate Std. Error z value Pr(>|z|)
     12 (Intercept) -1.149e+01  5.394e-03   -2131  <2e-16 ***
     13 lO.GRP       1.885e+00  4.306e-04    4376  <2e-16 ***
     14 1D.GRP       1.670e+00  4.401e-04    3794  <2e-16 ***
     15 1DistRd     -2.191e+00  7.909e-04   -2770  <2e-16 ***
     16 ---
     17 Signif. codes:  0 *** 0.001 ** 0.01 * 0.05 . 0.1   1
     18
     19 (Dispersion parameter for poisson family taken to be 1)
     20
     21      Null deviance: 45490237  on 991  degrees of freedom
     22 Residual deviance: 10260808  on 988  degrees of freedom
     23 AIC: 10270760
     24
     25 Number of Fisher Scoring iterations: 5
```

We see that effects due to origin, destination, and distance are all found to be highly significant, as would be expected from examination of the scatterplots in Fig. 9.2.

Nevertheless, a comparison of the Akaike information criterion (AIC) statistic[3] for the two models

---

[3]The AIC statistic for a likelihood-based model, with k-dimensional parameter η, is defined as $AIC = -2\ell(\hat{\eta}) + 2k$, where $\ell(\eta)$ is the log-likelihood evaluated at η, and $\hat{\eta}$ is the maximum likelihood estimate of η. This statistic, as with others of its type, provides an estimate of the generalization error associated with the fitted model, in this case effectively by off-setting the assessment of how well the model fits the data by a measure of its complexity. See, for example, Hastie, Tibshirani, and Friedman [6, Sect. 7.5] for additional details.

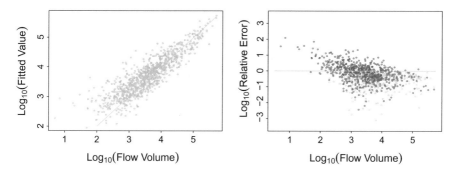

Fig. 9.3 Accuracy of estimates of traffic volume made by the gravity models for the Austrian call data. *Left*: Fitted values versus flow volume. *Right*: Relative error versus flow volume, where *light* and *dark* points indicate under- and over-estimation, respectively. All axes are on logarithmic scales, base ten. The lines $y = x$ and $y = 0$ are shown in *yellow* in the *left* and *right* plots, respectively, for reference

```
#9.10 1 > gm.g$aic
      2 [1] 5466814
      3 > gm.s$aic
      4 [1] 10270760
```

indicates that the more general model is vastly better than the simpler model, without necessarily overfitting, despite the fact that the former incorporates 64 variables, compared to four variables in the latter.

A comparison of the fitted values $\hat{\mu}_{ij}$, from the model gm.g, against the observed flow volumes z_{ij} is shown in Fig. 9.3.

```
#9.11 1 > plot(calldata$lFlowCnt,log(gm.g$fitted.values,10),
      2 +    cex.lab=1.5,
      3 +    xlab=expression(Log[10](paste("Flow Volume"))),
      4 +    col="green", cex.axis=1.5, ylab="", ylim=c(2, 5.75))
      5 > mtext(expression(Log[10](paste("Fitted Value"))), 2,
      6 +    outer=T, cex=1.5, padj=1)
      7 > clip(0.5,5.75,2,5.75)
      8 > abline(0, 1, lwd=2, col="darkgoldenrod1")
```

This comparison is shown on a log-log scale, due to the large dynamic range of the values involved. The relationship between the two quantities is found to be fairly linear, although arguably a bit better for medium- and large-volume flows than for low-volume flows.

Also shown in Fig. 9.3 is a comparison of the relative errors $\left(z_{ij} - \hat{\mu}_{ij}\right)/z_{ij}$ against the flow volumes z_{ij}. The comparison is again on a log-log scale. For the relative errors, the logarithm is applied to the absolute value, and then the sign of the error is reintroduced through the use of two shades of color.

```
#9.12 1 > res <- residuals.glm(gm.g, type="response")
      2 > relres <- res/calldata$FlowCnt
      3 > lrelres <- log(abs(relres),10)
```

```
 4 > res.sgn <- (relres>=0)
 5 >
 6 > plot(calldata$lFlowCnt[res.sgn], lrelres[res.sgn],
 7 +    xlim=c(0.5,5.75), ylim=c(-3.5,3.5),
 8 +    xlab=expression(Log[10](paste("Flow Volume"))),
 9 +    cex.lab=1.5, cex.axis=1.5, ylab="", col="lightgreen")
10 > mtext(expression(Log[10](paste("Relative Error"))), 2,
11 +    outer=T, cex=1.5, padj=1)
12 > par(new=T)
13 > plot(calldata$lFlowCnt[!res.sgn], lrelres[!res.sgn],
14 +    xlim=c(0.5,5.75), ylim=c(-3.5, 3.5),
15 +    xlab=expression(Log[10](paste("Flow Volume"))),
16 +    cex.lab=1.5, cex.axis=1.5, ylab="", col="darkgreen")
17 > mtext(expression(Log[10](paste("Relative Error"))), 2,
18 +    outer=T, cex=1.5, padj=1)
19 > clip(0.5,5.75,-3.5,3.5)
20 > abline(h=0, lwd=2, col="darkgoldenrod2")
```

We see that the relative error varies widely in magnitude. A large proportion of the flows are estimated with an error on the order of z_{ij} or less (i.e., the logarithm of relative error is less than zero), but a substantial number are estimated with an error on the order of up to ten times z_{ij}, and a few others are even worse. In addition, we can see that, roughly speaking, the relative error decreases with volume. Finally, it is clear that for low volumes the model is inclined to over-estimate, while for higher volumes, it is increasingly inclined to under-estimate.

9.3 Predicting Network Flows: Traffic Matrix Estimation

In many types of networks, it is difficult—if not effectively impossible—to actually measure the flow volumes Z_{ij} on a network. Nevertheless, knowledge of traffic flow volumes is fundamental to a variety of network-oriented tasks, such as traffic management, network provisioning, and planning for network growth. Fortunately, in many of the same contexts in which measurement of the flow volumes Z_{ij} between origins and destinations is difficult, it is often relatively easy to instead measure the flow volumes on network links. For example, in highway road networks, sensors may be positioned at the entrances to on- and off-ramps. Similarly, routers in an Internet network come equipped with the facility to monitor the data on incident links. Measurements of flow volumes on network links, in conjunction with knowledge of the manner in which traffic flows over these links, between origins and destinations, can be sufficient to allow us to accurately predict origin-destination volumes. This is known as the *traffic matrix estimation* problem.

Formally, let X_e denote the total flow over a given link $e \in E$, and let $\mathbf{X} = (X_e)_{e \in E}$. The link totals in $\mathbf{X}$ can be related to the origin-destination flow volumes in $\mathbf{Z}$ through the expression $\mathbf{X} = \mathbf{BZ}$, where $\mathbf{Z}$ now represents our traffic matrix written as a vector and $\mathbf{B}$ is the so-called *routing matrix*, describing the manner in which traffic moves throughout the network. The traffic matrix estimation problem then refers to predicting the Z_{ij} from the observed link counts $\mathbf{X} = (X_e)_{e \in E}$. Here we will focus on the case that each origin-destination pair (i, j) has only a single route between them, in which case $\mathbf{B}$ is a binary matrix,[4] with the entry in the row corresponding to link e and the column corresponding to pair (i, j) being

$$B_{e;ij} = \begin{cases} 1, & \text{if link } e \text{ is traversed in going from } i \text{ to } j, \\ 0, & \text{otherwise}. \end{cases} \qquad (9.11)$$

In any case of practical interest, the traffic matrix estimation problem will be highly under-constrained, in the sense that we effectively seek to invert the routing matrix $\mathbf{B}$ in the relation $\mathbf{X} \approx \mathbf{BZ}$, and $\mathbf{B}$ typically has many fewer rows (i.e., network links) than columns (i.e., origin-destination pairs). Various additional sources of information therefore typically are incorporated into the problem, which effectively serve to better constrain the set of possible solutions. Methods proposed in this area can be roughly categorized as *static* or *dynamic*, depending on whether they are aimed at estimating a traffic matrix for a single time period or successively over multiple time periods. In this section we will content ourselves, after a brief illustration of the nature of the traffic estimation problem, with introducing just one static method—the so-called tomogravity method—which has had a good deal of success in the context of the Internet.

9.3.1 An Ill-Posed Inverse Problem

The various static methods proposed for traffic matrix estimation are similar in that they all ultimately involve the optimization of some objective function, usually subject to certain constraints. However, they can differ widely in the construction of and justification for this objective function and, to a lesser extent, the constraints imposed.

One common approach is through the use of least-squares principles, which can be motivated by a Gaussian measurement model. This model specifies that

$$\mathbf{X} = \mathbf{B}\mu + \varepsilon, \qquad (9.12)$$

where $\mathbf{X} = (X_e)_{e \in E}$ and $\mathbf{B}$ are defined as above, μ is an $IJ \times 1$ vector of expected flow volumes over all origin destination pairs, and ε is an $N_e \times 1$ vector of errors. In principle, this formulation suggests that μ be estimated through ordinary least squares, i.e., by minimizing

$$(\mathbf{x} - \mathbf{B}\mu)^T (\mathbf{x} - \mathbf{B}\mu), \qquad (9.13)$$

---

[4]If multiple routes are possible, the entries of $\mathbf{B}$ are instead fractions representing, for example, the proportion of traffic from i to j that is expected to use the link e.

Fig. 9.4 Schematic
representation of the small
computer network
underlying the data in
`bell.labs`

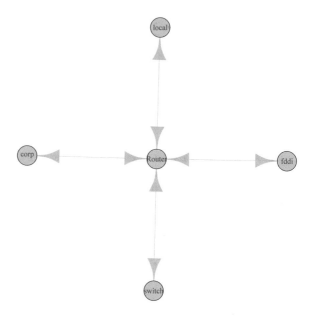

where **x** is the vector of observed link flow volumes. However, typically N_e is much smaller than IJ, and so the least-squares problem is under-determined and there will in general be an infinite number of possible solutions $\hat{\mu}$. That is, this traffic matrix estimation problem is ill-posed.

To illustrate, we will use the data set `bell.labs` in the **networkTomography** package.

```
#9.13 1 > library(networkTomography)
      2 > data(bell.labs)
```

These data correspond to measurements of traffic flow volume on a small computer network at Bell Laboratories, Lucent Technologies, as reported by Cao et al. [1]. The network consisted of four computing devices—named *fddi*, *local*, *switch*, and *corp*—connected to a single router. A representation of this network is shown in Fig. 9.4.

```
#9.14 1 > g.bl <- graph_from_literal(fddi:switch:local:corp
      2 +                              ++ Router)
      3 > plot(g.bl)
```

Renaming some of the variables in this data set in a manner consistent with the notation of this chapter,

```
#9.15 1 > B <- bell.labs$A
      2 > Z <- bell.labs$X
      3 > x <- bell.labs$Y
```

we see that we have available to us not only the link flow volumes **X** and the routing matrix **B**, but also the actual origin-destination flow volumes **Z**, the latter which are useful for purposes of validation. The flow volumes in **X** and **Z** are in units of average bytes per five-minute interval. The data were collected continuously over a twenty-four period.

Ignoring the single router in this network, which simply passes on traffic it receives, rather than itself generating traffic, there are a total of four vertices in this network with a total of eight links (i.e., one in-going and out-going link per vertex). Hence, we have 8 link-level traffic time series. Using the **lattice** package, and working from the data frame version of the underlying network and measurements encoded in bell.labs$df, we can produce a useful visualization that shows all eight of these time series simultaneously.

```
#9.16  1  > library(lattice)
       2  > traffic.in <- c("dst fddi","dst switch",
       3  +     "dst local","dst corp")
       4  > traffic.out <- c("src fddi","src switch",
       5  +     "src local","src corp")
       6  > my.df <- bell.labs$df
       7  > my.df$t <- unlist(lapply(my.df$time, function(x) {
       8  +     hrs <- as.numeric(substring(x, 11, 12))
       9  +     mins <- as.numeric(substring(x, 14, 15))
      10  +     t <- hrs + mins/60
      11  +     return(t)}))
      12  >
      13  > # Separate according to whether data
      14  > # are incoming or outgoing.
      15  > my.df.in <- subset(my.df, nme %in% traffic.in)
      16  > my.df.out <- subset(my.df, nme %in% traffic.out)
      17
      18  >
      19  > # Set up trellis plots for each case.
      20  > p.in <- xyplot(value / 2^10 ~ t | nme, data=my.df.in,
      21  +     type="l", col.line="goldenrod",
      22  +     lwd=2, layout=c(1,4),
      23  +     xlab="Hour of Day", ylab="Kbytes/sec")
      24  > p.out <- xyplot(value / 2^10 ~ t | nme, data=my.df.out,
      25  +     type="l", col.line="red",
      26  +     lwd=2, layout=c(1,4),
      27  +     xlab="Hour of Day", ylab="Kbytes/sec")
      28  >
      29  > # Generate trellis plots.
      30  > print(p.in, position=c(0,0.5,1,1), more=TRUE)
      31  > print(p.out, position=c(0,0,1,0.5))
```

The resulting plots are shown in Fig. 9.5. It is evident, looking at some of the more pronounced features of the individual curves, how the traffic from one source (e.g., *corp*, around hour 12) contributes to the traffic to another destination (e.g., *switch*, at the same time). The goal of traffic matrix estimation is to extract these underlying individual contributions from the aggregate link-level measurements shown in the figure.

Now let us examine the routing matrix for this network. The matrix B comes to us expressed in a full-rank form, by having dropped one row from the overall

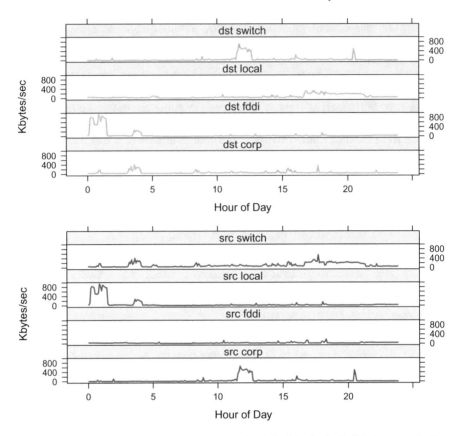

Fig. 9.5 Time series showing traffic volume over the eight links in the Bell Labs network. *Top*: Traffic by destination. *Bottom*: Traffic by source

routing matrix. For the purposes of illustration, we augment this matrix by adding back this missing row (yielding a matrix of reduced rank, but not changing the overall problem).

```
> B.full <- rbind(B, 2 - colSums(B))
> write.table(format(B.full),
+   row.names=F, col.names=F, quote=F)
1 1 1 1 0 0 0 0 0 0 0 0 0 0 0 0
0 0 0 0 1 1 1 1 0 0 0 0 0 0 0 0
0 0 0 0 0 0 0 0 1 1 1 1 0 0 0 0
0 0 0 0 0 0 0 0 0 0 0 0 1 1 1 1
1 0 0 0 1 0 0 0 1 0 0 0 1 0 0 0
0 1 0 0 0 1 0 0 0 1 0 0 0 1 0 0
0 0 1 0 0 0 1 0 0 0 1 0 0 0 1 0
0 0 0 1 0 0 0 1 0 0 0 1 0 0 0 1
```

This matrix has sixteen columns, corresponding to each origin-destination pair (including the four cases where a machine sends traffic to itself). Similarly, it has eight rows, corresponding to the four machines *fddi*, *local*, *switch*, and *corp*, respectively, first as origins and then as destinations. It is straightforward to derive this matrix from the topology of the network shown in Fig. 9.4. For example, the second column corresponds to the flow from *fddi* to *local*, while the seventh column corresponds to the flow from *local* to *switch*.

Clearly for this network the least squares problem in (9.13) is under-determined, in seeking to recover information on sixteen origin-destination flows from only eight link-level flows.

9.3.2 The Tomogravity Method

A standard approach to tackling ill-posed problems like that in (9.13) is to augment the objective function with some sort of penalty (also referred to as regularization), so as to restrict the collection of possible solutions in a useful manner. The *tomogravity method* is in this spirit, for the specific problem of traffic matrix estimation. Introduced by Zhang, Roughan, Lund, and Donoho [11], the 'tomo' part of the name refers to the fact that the traffic matrix estimation problem is considered a form of 'network tomography' (we encountered another example in Sect. 7.4), a term coined by Vardi [10]. The 'gravity' part of the name, in turn, refers to the fact that a simple gravity model is used to constrain the solutions produced by this method.

In the tomogravity method, we replace the least-squares criterion in (9.13) by a penalized least-squares criterion of the form

$$(\mathbf{x} - \mathbf{B}\mu)^T (\mathbf{x} - \mathbf{B}\mu) + \lambda D \left(\mu \,||\, \mu^{(0)}\right) \,, \tag{9.14}$$

where again the goal is to minimize the overall objective function. Here the penalty

$$D(\mu \,||\, \mu^{(0)}) = \sum_{ij} \frac{\mu_{ij}}{\mu_{++}} \log \frac{\mu_{ij}}{\mu_{ij}^{(0)}} \tag{9.15}$$

is the relative entropy 'distance' between a candidate solution μ and some pre-specified vector $\mu^{(0)}$. This quantity is also known as the Kullback–Liebler divergence and summarizes how similar the two 'distributions' $\{\mu_{ij}/\mu_{++}\}$ and $\{\mu_{ij}^{(0)}/\mu_{++}^{(0)}\}$ are to each other, where μ_{++} and $\mu_{++}^{(0)}$ are the sum of the elements in μ and $\mu^{(0)}$, respectively. It is always non-negative and will be equal to zero if and only if $\mu = \mu^{(0)}$.

The notion of a gravity model enters into the tomogravity method by specifying $\mu^{(0)}$ to have a certain multiplicative form reminiscent of a simple gravity model. Specifically, we let

$$\mu_{ij}^{(0)} = Z_{i+}^{(0)} \times Z_{+j}^{(0)} \tag{9.16}$$

be the product of the net out-flow and in-flow at vertices i and j, respectively. In addition, the constraint $\mu_{++} = Z_{++}^{(0)}$ is enforced, where $Z_{++}^{(0)}$ is the total traffic in the network. Importantly, these values all can be obtained through appropriate summation of the elements in the vector **x** of observed link counts.

Also necessary to completely specify the criterion in (9.14) is the value λ. This value acts as a smoothing parameter, with larger values encouraging solutions $\hat{\mu}$ that are closer to the pre-specified $\mu^{(0)}$. A value of $\lambda = (0.01)^2$ is report to work well in practice.

The regularized least-squares problem in (9.14) can be solved by methods of convex optimization. Note that $\hat{\mu}$, as an estimate of the expected flow volumes μ, does not necessarily satisfy the constraint $\mathbf{x} = \mathbf{Bz}$. If it is desired to enforce this constraint (i.e., if the measured link volumes **x** are without error), then the iterative proportional fitting procedure (IPFP) may be used. IPFP, usually credited to Deming and Stephan [4] in the statistics literature, is a standard tool in the statistical analysis of contingency tables, where it is used to force the elements of a table to match the marginals. See [1, Sect. 2.4] for a description of its use in the context of traffic matrix estimation.

The package **networkTomography** implements the tomogravity method, with IPFP adjustment, in the function `tomogravity`. We illustrate by applying it to the Bell Labs data.

```
#9.18  1 > x.full <- Z %*% t(B.full)
       2 > tomo.fit <- tomogravity(x.full, B.full, 0.01)
       3 > zhat <- tomo.fit$Xhat
```

The plotting capabilities of the **lattice** package are again useful, now for comparing the resulting predictions for the sixteen origin-destination pairs in this network against the actual origin-destination flows.

```
#9.19  1 > nt <- nrow(Z); nf <- ncol(Z)
       2 > t.dat <- data.frame(z = as.vector(c(Z) / 2^10),
       3 +     zhat = as.vector(c(zhat) / 2^10),
       4 +     t <- c(rep(as.vector(bell.labs$tvec), nf)))
       5 >
       6 > od.names <- c(rep("fddi->fddi", nt),
       7 +     rep("fddi->local", nt),
       8 +     rep("fddi->switch", nt), rep("fddi->corp",nt),
       9 +     rep("local->fddi", nt), rep("local->local",nt),
      10 +     rep("local->switch", nt), rep("local->corp",nt),
      11 +     rep("switch->fddi", nt), rep("switch->local",nt),
      12 +     rep("switch->switch", nt), rep("switch->corp",nt),
      13 +     rep("corp->fddi", nt), rep("corp->local",nt),
      14 +     rep("corp->switch", nt), rep("corp->corp",nt))
      15 >
      16 > t.dat <- transform(t.dat, OD = od.names)
      17 >
      18 > xyplot(z~t | OD, data=t.dat,
      19 +     panel=function(x, y, subscripts){
      20 +         panel.xyplot(x, y, type="l", col.line="blue")
      21 +         panel.xyplot(t.dat$t[subscripts],
      22 +                      t.dat$zhat[subscripts],
      23 +                      type="l", col.line="green")
```

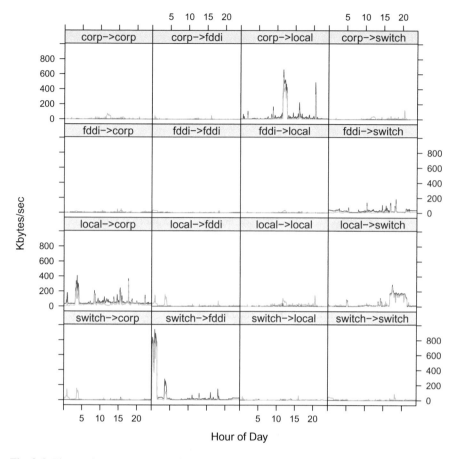

Fig. 9.6 Time series showing the prediction of the actual origin-destination flow volumes (*blue*) by the tomogravity method (*green*) for the Bell Labs network

```
24 +    }, as.table=T, subscripts=T, xlim=c(0,24),
25 +    xlab="Hour of Day", ylab="Kbytes/sec",
26 +    layout=c(4,4))
```

The results are shown in Fig. 9.6 and indicate that the tomogravity method appears to do quite well in capturing the flow volumes across the network, despite the highly variable nature of these time series and the somewhat substantial dynamic range (i.e., almost one megabyte order of magnitude on the *y*-axis).

9.4   Additional Reading

For a thorough introduction to the topic of gravity models, see Sen and Smith [8].
A review of the traffic matrix estimation problem, and other related problems in
network tomography, may be found in Castro et al. [3].

References

1. J. Cao, D. Davis, S. Wiel, and B. Yu, "Time-varying network tomography: router link data."
    Journal of the American Statistical Association, vol. 95, no. 452, pp. 1063–1075, 2000.
2. H. Carey, *Principles of Social Science*.   Philadelphia: Lippincott, 1858.
3. R. Castro, M. Coates, G. Liang, R. Nowak, and B. Yu, "Network tomography: recent develop-
    ments," *Statistical Science*, vol. 19, no. 3, pp. 499–517, 2004.
4. W. Deming and F. Stephan, "On a least squares adjustment of a sampled frequency table when
    the expected marginal totals are known," *Annals of Mathematical Statistics*, vol. 11, no. 4, pp.
    427–444, 1940.
5. M. Fischer and S. Gopal, "Artificial neural networks: a new approach to modeling interregional
    telecommunication flows," *Journal of Regional Science*, vol. 34, no. 4, pp. 503–527, 1994.
6. T. Hastie, R. Tibshirani, and J. Friedman, *The Elements of Statistical Learning*.   New York:
    Springer, 2001.
7. P. McCullagh and J. Nelder, *Generalized Linear Models*.   London: Chapman & Hall/CRC,
    1989.
8. A. Sen and T. Smith, *Gravity Models of Spatial Interaction Behavior*.   Berlin: Springer-Verlag,
    1995.
9. J. Stewart, "An inverse distance variation for certain social influences," *Science*, vol. 93, no.
    2404, pp. 89–90, 1941.
10. Y. Vardi, "Network tomography: estimating source-destination traffic intensities from link
    data," *Journal of the American Statistical Association*, vol. 91, no. 433, pp. 365–377, 1996.
11. Y. Zhang, M. Roughan, C. Lund, and D. Donoho, "An information-theoretic approach to traffic
    matrix estimation," in *Proceedings of SIGCOMM'03*, 2003.

Chapter 10
Networked Experiments

10.1 Introduction

Across the sciences—social, biological, and physical alike—there is a pervasive interest in evaluating the effect of treatments or interventions of various kinds. Generally, the ideal is understood to be to evaluate the proposed treatment in a manner unmarred by bias of any sort. On the other hand, nature and circumstances often conspire to make achievement of this ideal difficult (if not impossible). As a result, there is by now a vast literature on the design, conduct, and analysis of studies for evaluating the efficacy of treatment.

The prototypical example of such studies arguably is the randomized controlled trial [20], also called A/B testing more recently in the computer science literature [13]. It is by this that we will mean 'experiment' in this chapter. The defining characteristics of this experimental design are (i) the notion of two groups to be compared, i.e., treatment and control (or A and B) and (ii) the randomized assignment of individuals (e.g., people, cells, computers) to treatment and control groups. Ideally, these individuals will differ on average only in what treatment they receive. Importantly, also typical for these experiments is the assumption that the treatment received by any one individual does not affect the outcomes of any other individuals. That is, it is traditional to assume that there is no *interference* present in the experiment [6, p 19].

Increasingly, however, there is a marked interest in the assessment of treatment effects within networked systems. While experiments on networks are not new, recent advances in technology have in recent years facilitated a sea-change in both their scale and scope. Nowhere is this change more evident than in the social sciences, where experimental social science has undergone a phenomenal transformation, both within traditional areas like economics and sociology and within related areas like marketing. Leveraging the pervasiveness of social media platforms, so-called networked experiments in these areas have explored on previously unthinkable scales topics like the diffusion of knowledge and information, the ability of advertising to

© Springer Nature Switzerland AG 2020

E. D. Kolaczyk and G. Csárdi, *Statistical Analysis of Network Data with R*, Use R!,
https://doi.org/10.1007/978-3-030-44129-6_10

influence product adoption, and the spread of political opinions, to name just a few. For a relatively recent review of the literature in this dynamic area, see [3, Table 1].

Naturally, interference cannot realistically be assumed away when doing experiments on networks. As a result, much of what is considered standard in the traditional design of randomized experiments and the corresponding analysis for causal inference does not apply directly in this context. There has been a flurry of work in network analysis in recent years on better characterizing the problems posed by network-based interference (or, simply, 'network interference') and offering appropriate modifications to standard designs and methods of inference. In this chapter, we provide a brief but self-contained summary of one such modification. In Sect. 10.2 we present necessary notation, background, and illustrative context, specifically in regards to the notion of counterfactuals and the potential outcomes framework for causal inference. Then in Sect. 10.3 we address the topic of causal inference under network interference—first discussing network exposure models and causal estimands under interference, then commenting on design implications, and finally detailing procedures for estimation and quantification of uncertainty. We illustrate throughout with an example drawn from the context of organizational behavior.

10.2 Background

10.2.1 Causal Inference Using Potential Outcomes

Suppose that we have a finite population on which we would like to assess the effectiveness of a treatment, in comparison to the control condition of no treatment. In addition, suppose that there is some notion of a relation between pairs of individuals in our population that is expected to be relevant to the experiment. We represent this scenario through the use of a network graph $G = (V, E)$, where the vertices in V correspond to the N_v individuals in the population, and the edges in E, to their relationships. Note that the specification of such a G is effectively a modeling decision and depends on context. For example, in the context of the spread of a virus, an edge $\{i, j\} \in E$ might indicate some notion of 'contact' between individuals i and j during a certain period of time. Alternatively, the same edge might instead mean that individuals i and j are friends on Facebook in the context of an experiment assessing effectiveness of a certain advertising scheme in social media.

Let $z_i = 1$ indicate that individual $i \in V$ received the treatment. We will refer to $\mathbf{z} = (z_1, \ldots, z_{N_v})^T \in \{0, 1\}^{N_v}$ as the treatment assignment vector. By an experimental design we will mean a manner of choosing such a treatment vector. Let $p_{\mathbf{z}} = \mathbb{P}(\mathbf{Z} = \mathbf{z})$ be the probability that treatment assignment $\mathbf{z}$ is generated by the design. Finally, let $\mathcal{O}_i(\mathbf{z})$ denote the outcome for individual i under treatment $\mathbf{z}$. Our treatments here are binary and our outcomes will be assumed continuous, but extensions of what follows can be defined accordingly for other choices.

Table 10.1 Illustration of potential outcomes

Individual	'Ideal' World		Real World	
	$Z_i = 0$	$Z_i = 1$	$Z_i = 0$	$Z_i = 1$
1	$\mathcal{O}_1(0)$	$\mathcal{O}_1(1)$	$\mathcal{O}_1(0)$	?
2	$\mathcal{O}_2(0)$	$\mathcal{O}_2(1)$	?	$\mathcal{O}_2(1)$

The goal of evaluating treatment effectiveness based on the observed outcomes $\mathcal{O}_1(\mathbf{z}), \ldots, \mathcal{O}_{N_v}(\mathbf{z})$ is a problem of causal inference. As noted by Holland [9], the 'fundamental problem of causal inference' is the fact that, although we wish to assess the difference in outcomes of individuals under different treatment options, we are unable to measure multiple outcomes simultaneously on any given individual. The potential outcomes framework—also called the Rubin causal model, or sometimes the Neyman-Rubin causal model [1, 18]—is a framework for causal inference that approaches this problem by first making explicit the notion of all outcomes possible under the experimental design. That is, inclusive of both observed and counterfactual (potential) outcomes. This idea is illustrated in Table 10.1, for $N_v = 2$ individuals. In the ideal world, all four potential outcomes are available to us. However, in the real world, only two of those four are available.

Traditionally, in the standard version of this framework, it is assumed that there is no interference between individuals, i.e., that $\mathcal{O}_i(\mathbf{z}) = \mathcal{O}_i(z_i)$. This assumption is known in the potential outcomes literature as the stable unit treatment value assumption (SUTVA) and, in fact, was already implicit in our choice of notation in Table 10.1. Under this assumption it becomes meaningful to define the causal effect for individual i (traditionally called the unit-level causal effect) as a function solely of $\mathcal{O}_i(1)$ and $\mathcal{O}_i(0)$. The difference $\mathcal{O}_i(1) - \mathcal{O}_i(0)$ is a popular choice of definition, but in general other functions of these two variables (e.g., the ratio) can be used as well. In turn, various functions of these unit-level causal effects may be of interest in summarizing the effect of treatment at the level of the population in $V = \{1, \ldots, N_v\}$ or subpopulations thereof. A canonical causal estimand of interest is the average treatment effect

$$\tau_{ATE} = \frac{1}{N_v} \sum_{i=1}^{N_v} [\mathcal{O}_i(1) - \mathcal{O}_i(0)] = \bar{\mathcal{O}}(1) - \bar{\mathcal{O}}(0) . \tag{10.1}$$

Note that the estimand in (10.1) is a finite-sample population average. Here the perspective is one of design-based inference, in the sense that the randomness in the data we observe is assumed due entirely to the randomization in the experimental design.[1] This randomization is captured through the distribution $p_\mathbf{z}$ in our notation above. The standard agenda in this setting is typically as follows: define an unbiased

---

[1] In contrast, model-based inference adopts the perspective of the outcomes $\mathcal{O}_i(\cdot)$ as being drawn from a super-population, either due to measurement error on specific individuals i or random selection of individuals from a larger population—or both.

estimator $\hat{\tau}_{ATE}$ of τ_{ATE}, write down the variance $\mathrm{Var}(\hat{\tau}_{ATE})$ of that estimator, formulate an estimate $\widehat{Var}(\hat{\tau}_{ATE})$ of that variance, and quantify the uncertainty in $\hat{\tau}_{ATE}$ through, for example, confidence intervals. The details of this agenda depend on the form of p_z. Generally, however, while defining $\hat{\tau}_{ATE}$ often is not difficult, estimation of its variance frequently is nontrivial.

For example, consider the classic case of the completely randomized experiment, wherein N_t individuals are assigned to treatment, and $N_c = N_v - N_t$, to control, completely at random. Then $p_z = \binom{N_v}{N_t}^{-1}$ and an unbiased estimator of τ_{ATE} is just

$$\hat{\tau}_{ATE}(\mathbf{z}) = \frac{1}{N_v} \sum_{i=1}^{N_v} \left[\frac{z_i \mathcal{O}_i(1)}{N_t/N_v} - \frac{(1-z_i)\mathcal{O}_i(0)}{N_c/N_v} \right] . \tag{10.2}$$

The variance of this estimator is given by

$$\mathrm{Var}\left(\hat{\tau}_{ATE}(\mathbf{Z})\right) = \frac{S_c^2}{N_c} + \frac{S_t^2}{N_t} - \frac{S_{tc}^2}{N_v} , \tag{10.3}$$

where

$$S_c^2 = \frac{1}{N_v - 1} \sum_{i=1}^{N_v} \left[\mathcal{O}_i(0) - \bar{\mathcal{O}}(0)\right]^2 \tag{10.4}$$

and

$$S_t^2 = \frac{1}{N_v - 1} \sum_{i=1}^{N_v} \left[\mathcal{O}_i(1) - \bar{\mathcal{O}}(1)\right]^2 \tag{10.5}$$

are the population-level variances (using the frequent convention of $N - 1$, rather than N, for finite populations) of the individual potential outcomes under control and treatment, respectively, and, similarly,

$$S_{tc}^2 = \frac{1}{N_v - 1} \sum_{i=1}^{N_v} \left[\mathcal{O}_i(1) - \mathcal{O}_i(0) - \tau_{ATE}\right]^2 \tag{10.6}$$

is the variance of the treatment effects.

Note that the first two terms in (10.3) can be estimated in an unbiased fashion using the sample variances of the outcomes observed for control and treated groups, respectively. Furthermore, if the treatment effects $\mathcal{O}_i(1) - \mathcal{O}_i(0)$ are constant over all individuals i, then the third term in (10.3) is zero, and hence we have an unbiased estimate of $\mathrm{Var}(\hat{\tau}_{ATE})$. In general, however, we cannot estimate the variance in (10.6) based on the observed outcomes alone, since we do not observe both $\mathcal{O}_i(1)$ and $\mathcal{O}_i(0)$ for any of the individuals. As a result, common practice is to use the unbiased estimator derived under the assumption of constant treatment effect (expected to yield a conservative estimate of variance), or to derive estimators based on other or additional assumptions. Confidence intervals typically are based on asymptotic arguments. See [12, Ch 6], for example.

10.2.2 Network Interference: An Illustration

In the context of networks, we can in general expect that $\mathcal{O}_i(\mathbf{Z}) \neq \mathcal{O}_i(Z_i)$—that is, that there is interference—and so the stable unit value treatment assumption is not tenable. In fact, those managing an intervention may well be depending on there being interference.

Consider, for example, the strike dataset of Michael [15] (also analyzed in detail in [7, Chap. 7]).

```
#10.1  1 > library(sand)
       2 > data(strike)
```

New management took over at a forest products manufacturing facility, and this management team proposed certain changes to the compensation package of the workers. Two union negotiators among the workers were initially responsible for explaining the details of the proposed changes to the others. The changes were not accepted by the workers, and a strike ensued, which was then followed by a halt in negotiations. At the request of management, who felt that the information about their proposed changes was not being communicated adequately, an outside consultant analyzed the communication structure among 24 relevant workers. When the structure of the corresponding social network was revealed, two specific additional workers were approached and had the changes explained to them, which they then discussed with their colleagues. Within two days the workers requested that their union representatives re-open negotiations and the strike was resolved soon thereafter.

The network resulting from this process consists of 24 vertices and 38 edges.

```
#10.2  1 > summary(strike)
       2 IGRAPH 2669265 U--- 24 38 --
       3 + attr: names (v/c), race (v/c)
```

An edge between two vertices indicates that the corresponding employees communicated at some minimally sufficient level of frequency about the strike. Three subgroups are present in the network, i.e., older, English-speaking employees, younger, English-speaking employees, and younger, Spanish-speaking employees. These are encoded here through the vertex attribute race.

```
#10.3  1 > table(V(strike)$race)
       2 OE YE YS
       3 11  9  4
```

The strategy of explaining the proposed compensation package to a handful of actors in this network can be viewed as an intervention or treatment, in the language of this chapter. And, in hoping that the changes be discussed with other colleagues in the network, it is a treatment designed to exploit the phenomenon of network interference, with the goal of influencing organizational behavior. Since the ultimate goal of the treatment is to convince the employees to accept the proposed changes to the compensation package, the strategy can be viewed as a version of a network persuasion campaign (e.g., [4, 16]).

A visual representation of the strike network is useful. In creating this representation, we add another vertex shape option in **igraph**.[2]

```
#10.4  1  > # Create a triangle vertex shape
       2  > mytriangle <- function(coords, v=NULL, params) {
       3  +    vertex.color <- params("vertex", "color")
       4  +    if (length(vertex.color) != 1 && !is.null(v)) {
       5  +       vertex.color <- vertex.color[v]
       6  +    }
       7  +    vertex.size <- 1/200 * params("vertex", "size")
       8  +    if (length(vertex.size) != 1 && !is.null(v)) {
       9  +       vertex.size <- vertex.size[v]
      10  +    }
      11  +
      12  +    symbols(x=coords[,1], y=coords[,2], bg=vertex.color,
      13  +            stars=cbind(vertex.size, vertex.size, vertex.size),
      14  +            add=TRUE, inches=FALSE)
      15  + }
      16  > add_shape("triangle", clip=shapes("circle")$clip,
      17  +           plot=mytriangle)
```

Indicating race through vertex shape

```
#10.5  1  > V(strike)[V(strike)$race=="YS"]$shape <- "circle"
       2  > V(strike)[V(strike)$race=="YE"]$shape <- "square"
       3  > V(strike)[V(strike)$race=="OE"]$shape <- "triangle"
```

and distinguishing the four employee representatives using color,

```
#10.6  1  > nv <- vcount(strike)
       2  > z <- numeric(nv)
       3  > z[c(5,15,21,22)] <- 1
       4  > V(strike)$color <- rep("white",nv)
       5  > V(strike)[z==1]$color <- "red3"
```

our resulting visualization is shown in Fig. 10.1.

```
#10.7  1  > set.seed(42)
       2  > my.dist <- c(rep(1.8,4),rep(2.2,9),rep(2,11))
       3  > l <- layout_with_kk(strike)
       4  > plot(strike,layout=l,vertex.label=V(strike)$names,
       5  +      vertex.label.degree=-pi/3,
       6  +      vertex.label.dist=my.dist)
```

Note that, among the four employees in discussion with management,

```
#10.8  1  > V(strike)[z==1]$names
       2  [1] "Bob"      "Norm"      "Sam*"      "Wendle*"
```

---

[2]The default options are, essentially, circles, squares and rectangles. We introduce a triangle for greater visual distinction between the three vertex shapes we use to represent the variable race.

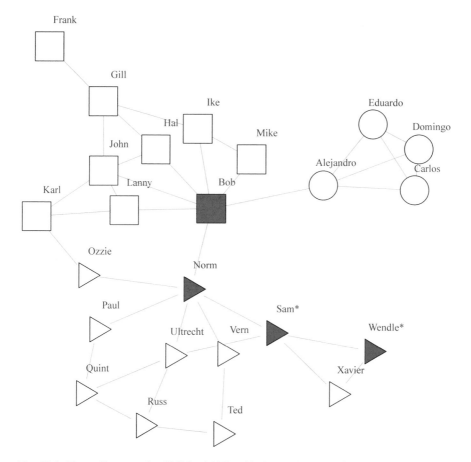

Fig. 10.1 The strike network of Michael [15], with three subgroups of employees indicated—younger, Spanish-speaking employees (circular vertices), younger, English-speaking employees (square vertices), and older, English-speaking employees (triangular vertices). Those actors serving as negotiators in the process are shown in red. The two union representatives—negotiators in the original communication attempt—are starred (i.e., Sam and Wendle)

whereas the two union representatives originally charged with explaining the new compensation package to the others (i.e., Sam and Wendle) seem relatively removed from the rest, the two employees who were later invited to join the conversations with management (i.e., Bob and Norm) appear to be much more central in the network. This visual impression can be confirmed, for example, by examining the relative rank of the four in terms of their betweenness and closeness centralities.

```
#10.9 1 > rank(-betweenness(strike))[z==1]
      2 [1]   1   2   4 21
      3 > rank(-closeness(strike))[z==1]
      4 [1]   1.0   2.0   6.5 22.0
```

While Bob and Norm rank first and second according to either measure, Sam lags behind them somewhat and Wendle ranks nearly last among the 24 employees in the network. This discrepency in turn has direct implications not only on the extent to which the desired messaging diffuses the network, but also on our ability to estimate this diffusion.

10.3   Causal Inference Under Network Interference

Much of the recent statistical work on networked experiments has focused upon extending the classical inference paradigm summarized in Sect. 10.2.1. In seeking to extend the classical potential outcomes framework to networked experiments, there is a natural tension between the complexity of the interference in the experiment and the corresponding impact that complexity has on both (i) defining meaningful estimands and (ii) producing estimators thereof and quantifying the uncertainty of those estimators. A natural approach to managing the challenges inherent in this complexity is to place assumptions upon the nature and extent of the interference. We focus here on the approach of Aronow and Samii [5], which does just that.

10.3.1   Network Exposure Models

In a networked experiment, while the experimenter (ideally) has control over assignment of treatments to individuals, the manner in which any given individual experiences a given treatment assignment as a whole is now assumed to be—at least in principle—a function of the networked system as a whole. As a result, the outcome for individual i arguably is better thought of as a result of the *exposure* of that individual to the full treatment assignment vector $\mathbf{z}$, rather than of only the specific treatment z_i to which that individual was assigned (as under SUTVA). But just how that network exposure manifests at the level of individuals is important. In the worst case, there will be 2^{N_v} possible exposures for each of the N_v individuals, making causal inference impossible.

Intuitively, to avoid this situation, modeling constraints are called for on the extent to which interference from other individuals in the network affect the exposure of a given individual i. Several approaches have been offered in the literature that formalize this idea. Aronow and Samii [5] introduce the notion of so-called exposure mappings, which we adopt here. The basic idea is to assume that effectively there are only some finite number K of conditions $\{c_1, \ldots, c_K\}$ to which any individual i is exposed. In this way, the complexity of possible exposures is thus reduced. Formally, we say that i is exposed to condition k if $f(\mathbf{z}, \mathbf{x}_i) = c_k$, where $\mathbf{z}$ is the treatment assignment vector, $\mathbf{x}_i$ is a vector of additional information specific to individual i, and f is the exposure mapping.

By way of illustration, note that in the classical setting, wherein it is assumed that there is no interference (i.e., SUTVA), the exposure mapping for each individual i is just $f(\mathbf{z}, \mathbf{x}_i) = z_i$, corresponding to just two conditions c_1 and c_2, i.e., treated or control. That is, exposure for individual i is determined entirely by the treatment assigned to that individual.

Alternatively, under interference, Aronow and Samii offer a simple, four-level categorization of exposure, which we will use as a running example. Let $\mathbf{A}$ denote the adjacency matrix of the network graph G, and take the vector $\mathbf{x}_i$ to be the i-th column of this matrix, i.e., $\mathbf{x}_i = \mathbf{A}_{\cdot i}$. Then define

$$f(\mathbf{z}, \mathbf{A}_{\cdot i}) = \begin{cases} c_{11} \text{ (Direct + Indirect Exposure)}, & z_i I_{\{\mathbf{z}^T \mathbf{A}_{\cdot i} > 0\}} = 1 \\ c_{10} \text{ (Isolated Direct Exposure)}, & z_i I_{\{\mathbf{z}^T \mathbf{A}_{\cdot i} = 0\}} = 1 \\ c_{01} \text{ (Indirect Exposure)}, & (1 - z_i) I_{\{\mathbf{z}^T \mathbf{A}_{\cdot i} > 0\}} = 1 \\ c_{00} \text{ (No Exposure)}, & (1 - z_i) I_{\{\mathbf{z}^T \mathbf{A}_{\cdot i} = 0\}} = 1 , \end{cases} \tag{10.7}$$

where the inner product $\mathbf{z}^T \mathbf{A}_{\cdot i}$ is equal to the number of directly exposed neighbors (or 'peers') of the vertex (or individual) i. Here a double subscript notation is used for the four exposure conditions to reflect the values of the two indicators z_i and $I_{\{\cdots\}}$. Note that under this definition, each individual i will fall into one and only one of these exposure conditions.

Under this particular model of exposure, in the strike network there are four employees who received both direct and indirect exposure,

```
#10.10  1 > A <- as_adjacency_matrix(strike)
        2 > I.ex.nbrs <- as.numeric(z%*%A > 0)
        3 > V(strike)[z*I.ex.nbrs==1]$names
        4 [1] "Bob"      "Norm"     "Sam*"     "Wendle*"
```

eleven employees who received only indirect exposure,

```
#10.11  1 > V(strike)[(1-z)*I.ex.nbrs==1]$names
        2 [1] "Alejandro" "Mike"     "Ike"       "Hal"
        3 [5] "John"      "Lanny"    "Ozzie"     "Paul"
        4 [9] "Vern"      "Xavier"   "Ultrecht"
```

and none who received isolated direct exposure.

```
#10.12  1 > V(strike)[z*(1-I.ex.nbrs)==1]$names
        2 character(0)
```

The rest of the employees received no exposure at all under this model.

```
#10.13  1 > V(strike)[(1-z)*(1-I.ex.nbrs)==1]$names
        2 [1] "Domingo" "Carlos"  "Eduardo" "Gill"    "Frank"
        3 [6] "Karl"    "Quint"   "Russ"    "Ted"
```

As we will see below, this choice of network exposure mapping allows for the definition of various causal estimands capturing to different extents the combination of direct and indirect effects of treatments.

10.3.2   Causal Effects Under Network Interference

With the addition of interference to the potential outcomes framework comes the necessity for a corresponding refinement of the notion of causal effects. Recall, for example, the classical definition of an average treatment effect τ_{ATE} in (10.1). A natural generalization of this quantity is

$$\frac{1}{N_v} \sum_{i=1}^{N_v} [\mathcal{O}_i(\mathbf{1}) - \mathcal{O}_i(\mathbf{0})] \;, \tag{10.8}$$

where $\mathbf{1}$ and $\mathbf{0}$ are now N_v-length vectors of ones and zeros, respectively. This notion of average treatment effect under network interference corresponds to a comparison of potential outcomes under full treatment (i.e., $\mathbf{z} = \mathbf{1}$) versus full control (i.e., $\mathbf{z} = \mathbf{0}$).[3] As such, it represents an aggregation of both direct and indirect effects of the treatment assignment. In many settings, however, it is of interest to disentangle these effects. For example, interference can be viewed alternately as a nuisance, with interest focused on extracting some notion of 'pure' treatment effects (such as in the context of a new putative drug therapy), or instead as the effect of central interest, often representing a mechanism to be exploited (such as from the perspective of social engineering). In the latter context, it is common to see terms like 'peer influence effects' and 'spillover effects' used in reference to portions of the treatment effect that are not isolated to the direct contribution of $z_i = 1$ to the outcome $\mathcal{O}_i(\mathbf{z})$.

In the general exposure mapping framework of Aronow and Samii, the approach to defining causal estimands is analogous to the classical case, but with the important difference that potential outcomes are defined now at the level of exposure conditions. Specifically, under this framework each individual i has K potential outcomes $\mathcal{O}_i(c_1), \ldots, \mathcal{O}_i(c_K)$. And it is assumed that each individual is exposed to one and only one condition. We use the difference $\mathcal{O}_i(c_k) - \mathcal{O}_i(c_l)$ to represent the causal effect for individual i of exposure condition k versus l. Then, in analogy to (10.1), define

$$\tau(c_k, c_l) = \frac{1}{N_v} \sum_{i=1}^{N_v} [\mathcal{O}_i(c_k) - \mathcal{O}_i(c_l)] \tag{10.9}$$

to be the average causal effect of exposure condition k versus l.

To use this framework for disentangling direct and indirect causal effects of treatment requires appropriate definition of $c_1, \ldots, c_K$. Consider again, for example, the exposure mapping function defined in (10.7). A natural set of contrasts are $\tau(c_{01}, c_{00})$, $\tau(c_{10}, c_{00})$, and $\tau(c_{11}, c_{00})$. The first contrast can be interpreted as capturing the average causal effect of treatment over control contributed to an untreated individual through treated neighbors. Note that this effect presumably includes not

---

[3] In fact, the expression in (10.8) is how average treatment effect is sometimes defined in the classical literature as well, but under the standard stable unit treatment value assumption this reduces to the commonly used expression in (10.1).

only that due to treatment of any of the neighbors themselves, but also that due to the treatment of the any neighbors' neighbors, and so on. Hence, it offers some notion of an overall indirect treatment effect. On the other hand, the second contrast is aimed at capturing some sense of the direct average causal effect of the treatment as applied to an individual i. That is, it is perhaps the closest in spirit to the classical average treatment effect τ_{ATE}. Finally, the third contrast effectively measures the average total treatment effect.

By way of illustration, in the strike network example, in order to mimic employee reaction to the new compensation plan proposed by management, we adopt a simple model in the spirit of the 'dilated effects' model of Rosenbaum [17]. Suppose that employee reaction is rated on a scale of 1 (non-receptive) to 10 (completely receptive). Let $\mathcal{O}_i(c_{00}) \equiv 1$ for all employees, indicating a common and standardized level of non-receptiveness to the proposal prior to any intervention. We then capture a sense of increased receptiveness as a function of increased exposure by specifying that $\mathcal{O}_i(c_{11}) = 10$, $\mathcal{O}_i(c_{10}) = 7$, and $\mathcal{O}_i(c_{01}) = 5$.

```
#10.14  1 > O.c11 <- 10.0; O.c10 <- 7; O.c01 <- 5; O.c00 <- 1.0
```

Thus, the effect of having the compensation package explained directly by management is to increase an employee's receptivity to the proposal from 1 to 7. But among those same employees, if they also are exposed to feedback from others like themselves, their receptivity goes up to a full 10. On the other hand, those employees hearing about the proposal only second-hand see a more moderate rise in receptivity to 5. The corresponding causal effects of interest are arguably $\tau(c_{11}, c_{00})$, $\tau(c_{10}, c_{00})$, and $\tau(c_{01}, c_{00})$, which respectively take the following values.

```
#10.15  1 > c(O.c11,O.c10,O.c01)-O.c00
        2 [1] 9 6 4
```

10.3.3 Design of Networked Experiments

Throughout our exposition so far in this chapter, the assignment of treatments has been left generic and unspecified. Ultimately, however, the choice of treatment assignment is a key element of network experimental design. At the most fundamental level, treatment assignment is captured in our framework simply through the distribution $p_{\mathbf{z}}$, i.e., the probability of treatment assignment $\mathbf{z}$. Yet under interference, as we have seen, it is more appropriate to think in terms of overall exposure of individuals to a treatment. Hence, for characterizing a network experimental design, it is useful to have a way of quantifying the manner in which treatment assignment induces exposure under a given network exposure model.

In the context of network exposure mappings, arguably the probability of most immediate interest is now the probability that an individual i is subject to exposure condition k, defined as

$$p_i^e(c_k) = \sum_{\mathbf{z}} p_{\mathbf{z}} \, I_{\{f(\mathbf{z}, \mathbf{x}_i) = c_k\}} \, . \tag{10.10}$$

Related probabilities of additional interest are the probability that individuals i and j are both subject to exposure k, i.e., $p_{ij}^e(c_k)$, and the probability that i and j are exposed to k and l, respectively, i.e., $p_{ij}^e(c_k, c_l)$. Formulas for these probabilities are analogous to that in (10.10). All three of these probabilities will be seen in the next section to play a key role in estimation and uncertainty quantification.

As an illustration, suppose that treatment is assigned to the N_v individuals in a network through Bernoulli random sampling, with probability p, and assume the network exposure model in (10.7). Then, for each individual i, there are four exposure probabilities:

$$
\begin{aligned}
p_i^e(c_{11}) &= p\left[1 - (1-p)^{d_i}\right] \\
p_i^e(c_{10}) &= p(1-p)^{d_i} \\
p_i^e(c_{01}) &= (1-p)\left[1 - (1-p)^{d_i}\right] \\
p_i^e(c_{00}) &= (1-p)^{d_i+1} \quad,
\end{aligned}
\tag{10.11}
$$

where d_i is the degree of vertex i in the network graph G. Closed-form expressions for the joint exposure probabilities can be obtained similarly but are more involved.

In general, for all individuals i and pairs of individuals i and j, the values of the various exposure probabilities can in principle be recovered from two classes of matrices. Suppose that there are M possible treatment assignments $\mathbf{z}$. Let

$$
\mathbf{I}_k =
\begin{bmatrix}
I_{\{f(\mathbf{z}_1,\mathbf{x}_1)=c_k\}} & I_{\{f(\mathbf{z}_2,\mathbf{x}_1)=c_k\}} & \cdots & I_{\{f(\mathbf{z}_M,\mathbf{x}_1)=c_k\}} \\
I_{\{f(\mathbf{z}_1,\mathbf{x}_2)=c_k\}} & I_{\{f(\mathbf{z}_2,\mathbf{x}_2)=c_k\}} & \cdots & I_{\{f(\mathbf{z}_M,\mathbf{x}_2)=c_k\}} \\
\vdots & \vdots & \ddots & \vdots \\
I_{\{f(\mathbf{z}_1,\mathbf{x}_{N_v})=c_k\}} & I_{\{f(\mathbf{z}_2,\mathbf{x}_{N_v})=c_k\}} & \cdots & I_{\{f(\mathbf{z}_M,\mathbf{x}_{N_v})=c_k\}}
\end{bmatrix} .
\tag{10.12}
$$

In addition, let $\mathbf{P} = \mathrm{diag}\left(p_{\mathbf{z}_1}, \ldots, p_{\mathbf{z}_M}\right)$. Then

$$
\mathbf{I}_k\mathbf{P}\mathbf{I}_k^T =
\begin{bmatrix}
p_1^e(c_k) & p_{12}^e(c_k) & \cdots & p_{1N_v}^e(c_k) \\
p_{21}^e(c_k) & p_2^e(c_k) & \cdots & p_{2N_v}^e(c_k) \\
\vdots & \vdots & \ddots & \\
p_{N_v1}^e(c_k) & p_{N_v2}^e(c_k) & \cdots & p_{N_v}^e(c_k)
\end{bmatrix}
\tag{10.13}
$$

and

$$
\mathbf{I}_k\mathbf{P}\mathbf{I}_l^T =
\begin{bmatrix}
0 & p_{12}^e(c_k, c_l) & \cdots & p_{1N_v}^e(c_k, c_l) \\
p_{21}^e(c_k, c_l) & 0 & \cdots & p_{2N_v}^e(c_k, c_l) \\
\vdots & \vdots & \ddots & \\
p_{N_v1}^e(c_k, c_l) & p_{N_v2}^e(c_k, c_l) & \cdots & 0
\end{bmatrix} .
\tag{10.14}
$$

The first matrix, in (10.13), is an $N_v \times N_v$ symmetric matrix, from which we can recover, for a fixed exposure condition c_k, both the individual exposure probabilities $p_i^e(c_k)$ and the joint exposure probabilities $p_{ij}^e(c_k)$, for all individuals

$i, j = 1, \ldots, N_v$. The second matrix, in (10.14), is also $N_v \times N_v$, but non-symmetric, and allows us to recover the joint exposure probabilities $p^e_{ij}(c_k, c_l)$ for all pairs of individuals i and j, for fixed exposure conditions c_k and c_l. Note that diagonal entries in (10.14) are all zero because of the assumption that individuals can only fall into one exposure category.

In practice, it is likely that (10.13) and (10.14) are computationally infeasible for direct calculation of $p^e_i(c_k)$, $p^e_{ij}(c_k)$, and $p^e_{ij}(c_k, c_l)$. However, Monte Carlo simulation can be used to approximate their values to arbitrary precision. Simulating n draws from p_z, for each of the K exposure conditions we can then form K matrices $\hat{\mathbf{I}}_k$ of dimension $N_v \times n$, in analogy to the definition of $\mathbf{I}_k$ in (10.12).

Suppose, for example, that in the strike network we 'treat' four employees chosen uniformly at random by naming them negotiators representing the others.

```
#10.16  1 > # Initialize
        2 > set.seed(41)
        3 > m <- 4   # Number of representatives
        4 > n <- 10000 # Number of Monte Carlo trials
        5 > I11 <- matrix(,nrow=nv,ncol=n)
        6 > I10 <- matrix(,nrow=nv,ncol=n)
        7 > I01 <- matrix(,nrow=nv,ncol=n)
        8 > I00 <- matrix(,nrow=nv,ncol=n)
        9 >
       10 > # Monte Carlo sampling
       11 > for(i in 1:n){
       12 +    z <- rep(0,nv)
       13 +    reps.ind <- sample((1:nv),m,replace=FALSE)
       14 +    z[reps.ind] <- 1
       15 +    reps.nbrs <- as.numeric(z%*%A > 0)
       16 +    I11[,i] <- z*reps.nbrs
       17 +    I10[,i] <- z*(1-reps.nbrs)
       18 +    I01[,i] <- (1-z)*reps.nbrs
       19 +    I00[,i] <- (1-z)*(1-reps.nbrs)
       20 + }
```

The estimators $\hat{\mathbf{I}}_k \hat{\mathbf{I}}_k^T / n$ and $\hat{\mathbf{I}}_k \hat{\mathbf{I}}_l^T / n$ are then unbiased for the matrices in (10.13) and (10.14), and converge almost surely by the strong law of large numbers.

```
#10.17  1 > I11.11 <- I11%*%t(I11)/n
        2 > I10.10 <- I10%*%t(I10)/n
        3 > I01.01 <- I01%*%t(I01)/n
        4 > I00.00 <- I00%*%t(I00)/n
```

These matrices can be visualized in the form of images. For example, the values for the exposure condition c_{00} can be visualized as follows.

```
#10.18  1  > # Plot c00
        2  > names.w.space <- paste(names," ",sep="")
        3  > my.cex.x <- 0.75; my.cex.y <- 0.75
        4  > image(I00.00, zlim=c(0,0.7), xaxt="n", yaxt="n",
        5  +       col=cm.colors(16))
        6  > mtext(side=1, text=names.w.space,at=seq(0.0,1.0,(1/23)),
        7  +       las=3, cex=my.cex.x)
        8  > mtext(side=2, text=names.w.space,at=seq(0.0,1.0,1/23),
        9  +       las=1, cex=my.cex.y)
       10  > mtext(side=3,text=expression("No Exposure"~(c["00"])),
       11  +       at=0.5, las=1)
       12  > # Add lines to differentiate groups.
       13  > u <- 1/23; uo2 <- 1/46
       14  > xmat <- cbind(rep(3*u+uo2,2),rep(12*u+uo2,2))
       15  > ymat <- cbind(c(0-uo2,1+uo2),c(0-uo2,1+uo2))
       16  > matlines(xmat,ymat, lty=1, lw=1, col="black")
       17  > matlines(ymat,xmat, lty=1, lw=1, col="black")
```

Values for the other three exposure conditions can be visualized similarly.

In Fig. 10.2 are shown the four resulting images. Not surprisingly, with only four actors out of 24 being 'treated' as negotiators, only the exposure probabilities for 'indirect exposure' and 'no exposure' have nontrivial values, both for individuals and pairs of individuals. Note that certain implications of network structure are visually apparent as well. For example, we see that in the subgroup of younger, Spanish-speaking employees there is a high probability of no exposure, both individually and pairwise. So too for certain subsets of the other two subgroups. Similarly, we see that Frank has especially high probability of having no exposure, and therefore joint exposure probabilities for this condition are correspondingly high for most other actors with Frank.

10.3.4   Inference for Causal Effects

Now consider the problem of inference for causal effects under network interference. Again, we focus on the exposure mapping framework, where estimands of fundamental interest are of two types: (i) the average potential outcomes $\bar{\mathcal{O}}(c_k)$, under exposure conditions c_k, for $k = 1, \ldots, K$, and (ii) the average causal effects $\tau(c_k, c_l) = \bar{\mathcal{O}}(c_k) - \bar{\mathcal{O}}(c_l)$ of exposure conditions k versus l, as defined in (10.9). Note that the observed outcomes corresponding to each estimand of the first type generally will be an unequal-probability sample without replacement of potential outcomes under a given condition, i.e., a sample from the set $\{\mathcal{O}_1(c_k), \ldots, \mathcal{O}_{N_v}(c_k)\}$ under condition c_k. Additionally, the estimands of the second type are all simple linear functions of those in the first. Therefore, assuming that we know or can approximate accurately the relevant exposure probabilities, it is natural to pursue inference following the method of Horvitz and Thompson [10].

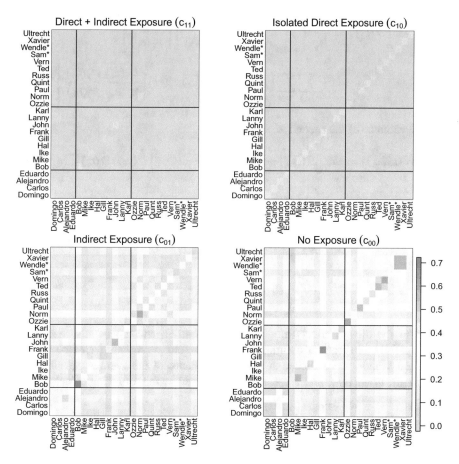

Fig. 10.2 Image-plot representations of inclusion probabilities $p_i^e(c)$ (diagonal entries) and $p_{ij}^e(c)$ (off-diagonal entries), as defined in (10.13), for the four exposure conditions defined in (10.7). Dark horizontal and vertical lines indicate subgroups

The Horvitz-Thompson framework accounts for unequal-probability sampling through the use of inverse probability weighting. Assume all individuals have non-zero exposure probabilities $p_i^e(c_k)$, for all exposure conditions c_k. Then the estimator

$$\hat{\bar{\mathcal{O}}}(c_k) = \frac{1}{N_v} \sum_{i=1}^{N_v} I_{\{f(\mathbf{z}, \mathbf{x}_i)\} = c_k\}} \frac{\mathcal{O}_i(c_k)}{p_i^e(c_k)} \tag{10.15}$$

is well-defined and unbiased for $\bar{\mathcal{O}}(c_k)$. In turn, $\hat{\tau}(c_k, c_l) = \hat{\bar{\mathcal{O}}}(c_k) - \hat{\bar{\mathcal{O}}}(c_l)$ is an unbiased estimator of $\tau(c_k, c_l)$. Furthermore, the variance of the first estimator is

$$\text{Var}\left[\hat{\bar{\mathscr{O}}}(c_k)\right] = \frac{1}{N_v^2}\left\{\sum_{i=1}^{N_v} p_i^e(c_k)\left[1 - p_i^e(c_k)\right]\left[\frac{\mathscr{O}_i(c_k)}{p_i^e(c_k)}\right]^2\right.$$

$$\left. + \sum_{i=1}^{N_v}\sum_{j\neq i}\left[p_{ij}^e(c_k) - p_i^e(c_k)p_j^e(c_k)\right]\frac{\mathscr{O}_i(c_k)}{p_i^e(c_k)}\frac{\mathscr{O}_j(c_k)}{p_j^e(c_k)}\right\} \quad (10.16)$$

and the variance of the second is

$$\text{Var}\left(\hat{\tau}(c_k, c_l)\right) = \text{Var}\left[\hat{\bar{\mathscr{O}}}(c_k)\right] + \text{Var}\left[\hat{\bar{\mathscr{O}}}(c_l)\right] - 2\text{Cov}\left[\hat{\bar{\mathscr{O}}}(c_k), \hat{\bar{\mathscr{O}}}(c_l)\right] , \quad (10.17)$$

where

$$\text{Cov}\left[\hat{\bar{\mathscr{O}}}(c_k), \hat{\bar{\mathscr{O}}}(c_l)\right] = \frac{1}{N_v^2}\left\{\sum_{i=1}^{N_v}\sum_{j\neq i}\frac{\mathscr{O}_i(c_k)}{p_i^e(c_k)}\frac{\mathscr{O}_j(c_l)}{p_j^e(c_l)}\left[p_{ij}^e(c_k, c_l) - p_i^e(c_k)p_j^e(c_l)\right]\right.$$

$$\left. - \sum_{i=1}^{N_v}\mathscr{O}_i(c_k)\mathscr{O}_j(c_l)\right\} . \quad (10.18)$$

Under the dilated effects model described earlier, and the (surely inaccurate, but hopefully illustrative) assumption that the four representatives Bob, Norm, Sam, and Wendle were chosen at random, estimates of the average causal effects $\tau(c_{11}, c_{00})$, $\tau(c_{10}, c_{00})$, and $\tau(c_{01}, c_{00})$ are obtained by the following calculations.

```
#10.19  1 > z <- rep(0,nv)
        2 > z[c(5,15,21,22)] <- 1
        3 > reps.nbrs <- as.numeric(z%*%A > 0)
        4 > c11 <- z*reps.nbrs
        5 > c10 <- z*(1-reps.nbrs)
        6 > c01 <- (1-z)*reps.nbrs
        7 > c00 <- (1-z)*(1-reps.nbrs)
        8 > Obar.c11 <- O.c11*mean(c11/diag(I11.11))
        9 > Obar.c10 <- O.c10*mean(c10/diag(I10.10))
       10 > Obar.c01 <- O.c01*mean(c01/diag(I01.01))
       11 > Obar.c00 <- O.c00*mean(c00/diag(I00.00))
       12 > print(c(Obar.c11,Obar.c10,Obar.c01)-Obar.c00)
       13 [1] 22.4876007 -0.8031629  5.9731170
```

These values compare rather poorly to their respective target values of 9, 6, and 4, which suggests that—despite being unbiased—the variance of these estimators is likely a concern.

As in the classical potential outcomes framework, where interference is absent, so too here the problem of variance estimation must be handled with care. For example, if the joint exposure probabilities $p_{ij}^e(c_k)$ are all strictly greater than zero, the estimator

$$\widehat{\text{Var}}\left[\hat{\bar{\mathscr{O}}}(c_k)\right] = \frac{1}{N_v^2}\left\{\sum_{i=1}^{N_v} I_{\{f(\mathbf{z},\mathbf{x}_i)=c_k\}}\left[1 - p_i^e(c_k)\right]\left[\frac{\mathscr{O}_i(c_k)}{p_i^e(c_k)}\right]^2\right.$$

$$+ \sum_{i=1}^{N_v}\sum_{j\neq i} I_{\{f(\mathbf{z},\mathbf{x}_i)=c_k\}}I_{\{f(\mathbf{z},\mathbf{x}_j)=c_k\}} \tag{10.19}$$

$$\left. \times \frac{p_{ij}^e(c_k) - p_i^e(c_k)p_j^e(c_k)}{p_{ij}^e(c_k)}\frac{\mathscr{O}_i(c_k)}{p_i^e(c_k)}\frac{\mathscr{O}_j(c_k)}{p_j^e(c_k)}\right\} \tag{10.20}$$

is an unbiased estimate of the variance in (10.16). If some of the joint exposure probabilities are equal to zero, then this variance estimator will be biased. However, its bias can be characterized. See Aronow and Samii [5, Sect. 5] for details, where a proposal for bias correction may also be found.

In contrast, note that the variance in (10.17) cannot be estimated in an unbiased or consistent manner, due to the last term for the covariance in expression (10.18), which depends upon (unobserved) potential outcomes. However, it is possible to produce a conservatively biased estimator of this variance, in the sense that its expectation is an upper bound on (10.17). See [5, Sect. 5].[4]

Here we use simulation to gain some insight into the performance of these estimators in the context of the strike network example.

```
#10.20  1 > set.seed(41)
        2 > n <- 10000
        3 > Obar.c11 <- numeric()
        4 > Obar.c10 <- numeric()
        5 > Obar.c01 <- numeric()
        6 > Obar.c00 <- numeric()
        7 > for(i in 1:n){
        8 +     z <- rep(0,nv)
        9 +     reps.ind <- sample((1:nv),m,replace=FALSE)
       10 +     z[reps.ind] <- 1
       11 +     reps.nbrs <- as.numeric(z%*%A > 0)
       12 +     c11 <- z*reps.nbrs
       13 +     c10 <- z*(1-reps.nbrs)
       14 +     c01 <- (1-z)*reps.nbrs
       15 +     c00 <- (1-z)*(1-reps.nbrs)
       16 +     Obar.c11 <- c(Obar.c11, O.c11*mean(c11/diag(I11.11)))
       17 +     Obar.c10 <- c(Obar.c10, O.c10*mean(c10/diag(I10.10)))
       18 +     Obar.c01 <- c(Obar.c01, O.c01*mean(c01/diag(I01.01)))
       19 +     Obar.c00 <- c(Obar.c00, O.c00*mean(c00/diag(I00.00)))
       20 + .}
       21 > ACE <- list(Obar.c11-Obar.c00, Obar.c10-Obar.c00,
       22 +             Obar.c01-Obar.c00)
```

We see that, as expected, the estimators of the three average causal effects $\tau(c_{11}, c_{00})$, $\tau(c_{10}, c_{00})$, and $\tau(c_{01}, c_{00})$ are all roughly zero.[5]

---

[4]Additionally, as in the classical potential outcomes framework, work to date for producing confidence intervals has rested on asymptotic arguments. See [5, Sect. 6].

[5]A larger number of trials n improves the approximation.

```
#10.21  1 > print(sapply(ACE,mean)-
        2 +          c(O.c11-O.c00, O.c10-O.c00, O.c01-O.c00))
        3 [1]   0.10811880 -0.02446790 -0.03481623
```

However, the corresponding standard errors are rather disparate.

```
#10.22  1 > print(sapply(ACE,sd))
        2 [1] 9.109026 3.736030 1.587795
```

More specifically, the coefficients of variation[6]

```
#10.23  1 > sapply(ACE,sd)/c(9,6,4)
        2 [1] 1.0121140 0.6226717 0.3969488
```

indicate that—relative to each other—our estimators are best at capturing $\tau(c_{01}, c_{00})$, and then $\tau(c_{10}, c_{00})$, and finally $\tau(c_{11}, c_{00})$. This ordering makes sense in light of the variance formula (10.17) and the differing orders of magnitude of the relevant inclusion probabilities (i.e., see Fig. 10.2).

10.4   Additional Reading

General background on the potential outcomes framework for causal inference may be found in the book by Imbens and Rubin [12]. The work of Halloran and Struchiner [8] is seminal in framing the nature of the complications that can arise under interference and in defining causal effects that attempt to disentangle direct and indirect treatment effects. Key contributions to the corresponding inference problems in this context include Sobel [19], Rosenbaum [2], Hudgens and Halloran [11], and Tchetgen-Tchetgen and VanderWeele [21]. For an overview of related literature from the econometrics perspective, see the review by Leung [14]. Finally, for a summary of some of the many examples of network experimental design used in practice, see the survey article by Aral [3], for example, and particularly Table 2 therein.

References

1. D. B. Rubin, "Comment: Neyman (1923) and causal inference in experiments and observa-tional studies," *Statistical Science*, Citeseer, vol. 5, no. 4, pp. 472–480, 1990.
2. P. R. Rosenbaum, "Interference between units in randomized experiments," *Journal of the American Statistical Association*, Taylor & Francis, vol. 102, no. 477, pp. 191–200, 2007.
3. S. Aral, "Networked experiments: A review of methods and innovations," in *The Oxford Hand-book of the Economics of Networks*, Y. Bramoulle, A. Galeotti, and B. Rogers, Eds. Oxford: Oxford University Press, 2016.
4. S. Aral and D. Walker, "Creating social contagion through viral product design: A randomized trial of peer influence in networks," *Management science*, vol. 57, no. 9, pp. 1623–1639, 2011.

---

[6]A measure of relative standard deviation, defined as the ratio of the standard deviation to the mean.

5. P. M. Aronow, C. Samii *et al.*, "Estimating average causal effects under general interference, with application to a social network experiment," *The Annals of Applied Statistics*, vol. 11, no. 4, pp. 1912–1947, 2017.

6. D. R. Cox, *Planning of Experiments*. Wiley, 1958.

7. W. De Nooy, A. Mrvar, and V. Batagelj, *Exploratory Social Network Analysis with Pajek*. Cambridge University Press, 2011, vol. 27.

8. M. Halloran and C. Struchiner, "Causal inference in infectious diseases," *Epidemiology*, vol. 6, no. 2, pp. 142–151, 1995.

9. P. W. Holland, "Statistics and causal inference," *Journal of the American Statistical Association*, vol. 81, no. 396, pp. 945–960, 1986.

10. D. Horvitz and D. Thompson, "A generalization of sampling without replacement from a finite universe," *Journal of the American Statistical Association*, vol. 47, no. 260, pp. 663–685, 1952.

11. M. Hudgens and M. Halloran, "Toward causal inference with interference," *Journal of the American Statistical Association*, vol. 103, no. 482, pp. 832–842, 2012.

12. G. W. Imbens and D. B. Rubin, *Causal Inference in Statistics, Social, and Biomedical Sciences*. Cambridge University Press, 2015.

13. R. Kohavi and R. Longbotham, "Online controlled experiments and a/b tests," *Encyclopedia of Machine Learning and Data Mining, C. Sammut and G. Webb, Eds*, 2015.

14. M. P. Leung, "Treatment and spillover effects under network interference," *Review of Economics and Statistics*, pp. 1–42, 2016.

15. J. H. Michael, "Labor dispute reconciliation in a forest products manufacturing facility," *Forest Products Journal*, vol. 47, no. 11/12, p. 41, 1997.

16. E. L. Paluck, "Peer pressure against prejudice: A high school field experiment examining social network change," *Journal of Experimental Social Psychology*, vol. 47, no. 2, pp. 350–358, 2011.

17. P. Rosenbaum, "Reduced sensitivity to hidden bias at upper quantiles in observational studies with dilated treatment effects," *Biometrics*, vol. 5, no. 2, pp. 560–564, 1999.

18. D. B. Rubin, "Estimating causal effects of treatments in randomized and nonrandomized studies." *Journal of Educational Psychology*, vol. 66, no. 5, p. 688, 1974.

19. M. Sobel, "What do randomized studies of housing mobility demonstrate? causal inference in the face of interference," *Journal of the American Statistical Association*, vol. 101, no. 476, pp. 1398–1407, 2006.

20. P. Solomon, M. M. Cavanaugh, and J. Draine, *Randomized Controlled Trials: Design and Implementation for Community-based Psychosocial Interventions*. Oxford University Press, 2009.

21. E. Tchetgen and T. VanderWeele, "On causal inference in the presence of interference," *Statistical Methods in Medical Research*, vol. 21, no. 1, pp. 55–75, 2012.

Chapter 11
Dynamic Networks

11.1 Introduction

Most complex systems—and, hence, networks—are dynamic in nature. So, realistically, the corresponding network graphs and processes thereon are dynamic as well and, ideally, should be analyzed as such. Friendships (both traditional and on-line versions) form and dissolve over time. Certain genes may regulate other genes, but only during specific stages of the natural cycle of a cell. And both the physical and logical structure of the Internet have been evolving ever since it was first constructed.

In practice, however, the vast majority of network analyses performed to date have been static. Arguably this state of affairs is due to a number of factors. First of all, broadly speaking, the methodology for dynamic network analysis is decidedly less developed than that for static network analysis. At the same time, unless handled carefully (and, sometimes, even then!), the computational burden associated with the analysis of dynamic network data can be comparatively much heavier. Additionally, it is not necessarily always the case that a complex system is varying in ways interesting or relevant to a study at the time scales upon which dynamics can be measured. That is, oftentimes a handful of well-performed static network analyses can sufficiently convey the behavior of the underlying system across time. And finally, sometimes it simply is not realistic to collect data indexed over time. For example, in systems biology it generally is far easier to collect measurements on gene transcription activity in a single, steady-state setting than at consecutive time points, particularly if temporal calibration across multiple cells is required.

Accordingly, the focus of this book has been devoted almost entirely to statistical methods for static network data. Nevertheless, given the importance of dynamic network data, and the expectation that the development of computationally efficient methods for its analysis will continue to receive increasingly greater attention in the coming years, we present here in this chapter a brief introduction to this quickly evolving area. The organization of the sections in this chapter mimics that of the earlier

© Springer Nature Switzerland AG 2020

E. D. Kolaczyk and G. Csárdi, *Statistical Analysis of Network Data with R*, Use R!,
https://doi.org/10.1007/978-3-030-44129-6_11

chapters in this book. Specifically, we discuss the representation and manipulation of dynamic network data in Sect. 11.2, its visualization in Sect. 11.3, its characterization in Sect. 11.4, and finally its modeling in Sect. 11.5.

11.2 Representation and Manipulation of Dynamic Networks

The adjective 'dynamic' has been applied in the context of networks to describe at least two different aspects of an evolving complex system. Most commonly, it is applied when the edges among a set of vertices—and sometimes the set of vertices itself—are changing as a function of time. Alternatively, it is sometimes used in reference to the attributes of the vertices or edges in a fixed graph G changing in time, such as in the case of the dynamic network processes discussed in Sect. 8.5 (e.g., epidemics). The former case may be thought of as referring to dynamics *of* a network, and the latter, to dynamics *on* a network. Of course, both types of dynamics may be present together—networks of this type sometimes are referred to as *co-evolving*. In this chapter our focus will be primarily on the first of the above cases.

Conceptually, we will think of a *dynamic network* as a time-indexed graph $G(t) = (V(t), E(t))$, with time t varying in either a discrete or continuous manner over some range of values. Here $V(t)$ is the set of vertices present in the network at time t, and $E(t)$, the set values $e_{ij}(t)$, for $i, j \in V(t)$, indicating the presence or absence of edges between those vertices. Sometimes it is convenient to let V be the collection of all vertices present at some point or another during the period of time under study, in which case we can write $V(t) \equiv V$ for all t.

In practice, the extent to which the time-varying behavior of a dynamic network $G(t)$ actually may be captured through measurements can vary. At one extreme, we may be able observe with complete accuracy the appearance and disappearance of each vertex in time, and similarly the formation and dissolution of edges between the vertices. That is, we may literally observe $G(t)$. And, of course, at the other extreme, we may only be able to obtain a marginal summary of those vertices and edges that appeared at any point during the period of observation, resulting in a static graph G. In between these two extremes lie various other possibilities. For example, we may observe a set of static snapshots of the network, each summarizing the marginal behavior of the network during successive time periods. Such data are sometimes referred to as panel data or longitudinal network data. Alternatively, we may have knowledge of interactions between vertices only up to some finite resolution in time.

An example of this last type of data is found in the data set hc.

```
#11.1 1 > library(sand)
       2 > data(hc)
```

These data,[1] first reported by Vanhems et al. [6], contain records of contacts among patients and various types of health care workers in the geriatric unit of a hospital in Lyon, France, in 2010, from 1pm on Monday, December 6 to 2pm on Friday, December 10. Each of the 75 people in this study consented to wear RFID sensors on small identification badges during this period, which made it possible to record when any two of them were in face-to-face contact with each other (i.e., within 1–1.5 m of each other) during a 20-s interval of time. A primary goal of this study was to gain insight into the pattern of contacts in such a hospital environment, particularly with an eye towards the manner in which infection might be transmitted.

The object hc is a data frame in which the measurements are stored as an edge list.

```
#11.2  1 > head(hc)
       2   Time ID1 ID2  S1   S2
       3 1  140  15   31 MED ADM
       4 2  160  15   22 MED MED
       5 3  500  15   16 MED MED
       6 4  520  15   16 MED MED
       7 5  560  16   22 MED MED
       8 6  580  16   22 MED MED
```

For each contact during the week-long period of observation there is a row, in which are recorded the time (in seconds) at which the corresponding 20-s interval terminated, an identification number for each of the two people involved in the contact, and the statuses of these people. Status was assigned according to one of four categories, i.e., administrative staff (ADM), medical doctor (MED), paramedical staff, such as nurses or nurses' aides (NUR), and patients (PAT). We note that the designation of '1' and '2' in the ID and status variables is arbitrary.

From a network-based perspective, we view the people in this study as vertices, and their status, as a vertex attribute. In order to get a sense of the distribution of this attribute across vertices, we merge the vertex IDs and status variables, extract the unique pairings of ID and status, and summarize the results in a table.

```
#11.3  1 > ID.stack <- c(hc$ID1,hc$ID2)
       2 > Status.stack <- c(as.character(hc$S1),
       3 +    as.character(hc$S2))
       4 > my.t <- table(ID.stack,Status.stack)
       5 > v.status <- character(nrow(my.t))
       6 > for(i in (1:length(v.status))){
       7 +    v.status[i] <- names(which(my.t[i,]!=0))
       8 + }
       9 > table(v.status)
      10 v.status
      11 ADM MED NUR PAT
      12   8  11  27  29
```

---

[1] Original data are available at http://www.sociopatterns.org.

We see that patients and nurses each make up roughly 40% of the study, while the administrators and medical doctors each make up roughly only 10%.

Similarly, we can view each contact between two people as an edge. Summarizing these contacts in a two-way table allows us to get some rough preliminary sense as to the distribution of such edges throughout our network.

```
#11.4  1 > status.t <- table(hc$S1,hc$S2)
       2 > status.t <- status.t + t(status.t)
       3 > diag(status.t) <- round(diag(status.t)/2)
       4 > status.t
       5        ADM    MED    NUR    PAT
       6  ADM   279    459   2596    441
       7  MED   459   5660   1769   1471
       8  NUR  2596   1769  12695   6845
       9  PAT   441   1471   6845    209
```

We see, for example, that easily the largest number of contacts was between nurses.

A more refined sense of the pattern of contacts among these individuals may be obtained by incorporating time into our analysis. Figure 11.1 shows histograms of these contacts, as a function of time, grouped by the status of the interacting pairs.

```
#11.5  1 > tmp.es <- paste(hc$S1,"-",hc$S2,sep="")
       2 > e.status <- character(dim(hc)[[1]])
       3 > e.status[tmp.es=="ADM-ADM"] <- "ADM-ADM"
       4 > e.status[tmp.es=="MED-MED"] <- "MED-MED"
       5 > e.status[tmp.es=="NUR-NUR"] <- "NUR-NUR"
       6 > e.status[tmp.es=="PAT-PAT"] <- "PAT-PAT"
       7 > e.status[(tmp.es=="ADM-MED") |
       8 +   (tmp.es=="MED-ADM")] <- "ADM-MED"
       9 > e.status[(tmp.es=="ADM-NUR") |
      10 +   (tmp.es=="NUR-ADM")] <- "ADM-NUR"
      11 > e.status[(tmp.es=="ADM-PAT") |
      12 +   (tmp.es=="PAT-ADM")] <- "ADM-PAT"
      13 > e.status[(tmp.es=="MED-NUR") |
      14 +   (tmp.es=="NUR-MED")] <- "MED-NUR"
      15 > e.status[(tmp.es=="MED-PAT") |
      16 +   (tmp.es=="PAT-MED")] <- "MED-PAT"
      17 > e.status[(tmp.es=="NUR-PAT") |
      18 +   (tmp.es=="PAT-NUR")] <- "NUR-PAT"
      19 >
      20 > my.hc <- data.frame(Time = hc$Time/(60*60),
      21 +                     ID1 = hc$ID1,
      22 +                     ID2 = hc$ID2,
      23 +                     Status = e.status)
      24 >
      25 > library(lattice)
      26 > histogram(~Time|Status, data=my.hc, xlab="Hours",
      27 +   layout=c(5,2))
```

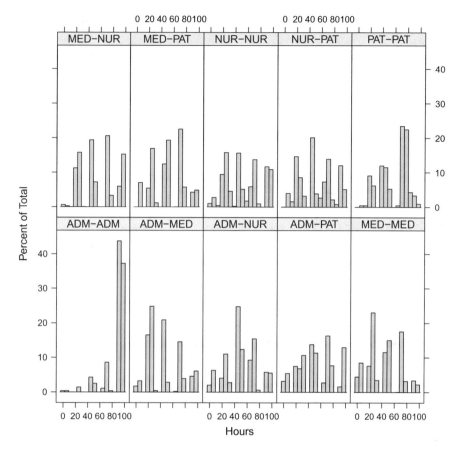

Fig. 11.1 Histograms of contacts in the hospital data set hc, grouped by the status of the interacting pair

Several interesting features now become apparent. For example, there is a clear diurnal structure to most of the histograms, with the vast majority of the contacts—not surprisingly—having occurred during the daytime or early evening hours. There are also exceptions to this pattern, the most notable of which may be the concentration of contacts between pairs of administrators (i.e., ADM–ADM) on the last day, suggesting the likelihood of a weekly organizational meeting.

Note that so far the notion of a network has only been implicit in our discussion. There are in fact several ways in which such data may be provided with an explicit network-based representation. One straightforward approach is through the use of multi-graphs, where each contact is represented by an edge, and each edge is equipped with an attribute indicating the time interval associated with it. This representation

can be implemented in a straightforward fashion in **igraph**. Applying it to our hospital
data

```
#11.6 1 > vids <- sort(unique(c(hc$ID1, hc$ID2)))
      2 > g.week <- graph_from_data_frame(hc[, c("ID1", "ID2",
      3 +     "Time")], vertices=data.frame(vids),
      4 +     directed=FALSE)
      5 > E(g.week)$Time <- E(g.week)$Time   / (60 * 60)
```

the resulting network is a single graph of 32, 424 (multi)edges, annotated with time
information, among 75 vertices.

```
#11.7 1 > summary(g.week)
      2 IGRAPH ab0ca23 UN-- 75 32424 --
      3 + attr: name (v/c), Time (e/n)
```

We also incorporate status as a vertex attribute.

```
#11.8 1 > status <- unique(rbind(data.frame(id=hc$ID1,
      2 +     status=hc$S1), data.frame(id=hc$ID2, status=hc$S2)))
      3 > V(g.week)$Status <-
      4 +     as.character(status[order(status[,1]),2])
```

Alternatively, we can represent such data in the form of a simple graph, by collapsing the multiple edges between vertex pairs to a single edge, but preserving
information on the multiplicity of edges through the use of edge weights. We do so
for the hospital data.

```
#11.9 1 > E(g.week)$weight <- 1
      2 > g.week.wgtd <- simplify(g.week)
      3 > summary(g.week.wgtd)
      4 IGRAPH 09000d9 UNW- 75 1139 --
      5 + attr: name (v/c), Status (v/c), weight (e/n)
```

We see that the resulting graph is indeed simple

```
#11.10 1 > is_simple(g.week.wgtd)
       2 [1] TRUE
```

and that its 1, 139 edges have weights varying over a number of orders of magnitude,
although the vast majority are on the order of 1–20.

```
#11.11 1 > summary(E(g.week.wgtd)$weight)
       2     Min. 1st Qu.  Median    Mean 3rd Qu.     Max.
       3     1.00    3.00    8.00   28.47   23.00 1059.00
```

Note that the time information is completely lost in collapsing a multi-graph to a
weighted graph. A compromise is to create 'slices' of the multi-graph, and collapse
within each slice. For example, we might take the first four days (96 h) of our hospital
data

```
#11.12 1 > g.week.96 <- subgraph.edges(g.week,
       2 +     E(g.week)[Time <= 96])
```

and separate that into eight slices of 12 h each.

```
#11.13 1 > g.sl12 <- lapply(1:8, function(i) {
       2 +    g <- subgraph.edges(g.week,
       3 +                         E(g.week)[Time > 12*(i-1) &
       4 +                                   Time <= 12*i],
       5 +                         delete.vertices=FALSE)
       6 +    simplify(g)
       7 + })
```

The resulting eight subnetworks are stored as a list of graphs. Each of these graphs has been defined to include the full set of 75 people as its set of vertices[2]

```
#11.14 1 > sapply(g.sl12,vcount)
       2 [1] 75 75 75 75 75 75 75 75
```

and is seen to contain on the order of two to three hundred pairs of vertices for whom the corresponding people were involved in at least one contact.

```
#11.15 1 > sapply(g.sl12,ecount)
       2 [1] 179 294 257 282 265 314 197 305
```

Finally, for measurements with sufficiently high temporal resolution—that is, for when we effectively observe $G(t) = (V(t), E(t))$ itself—it is natural to wish to maintain this representation of the dynamic network. The package **networkDynamic** allows for such a representation.

```
#11.16 1 > library(networkDynamic)
```

Specifically, each period during which a dynamic edge $e_{ij}(t)$ is 'on'—what the authors of this package refer to as a 'spell'—is represented by its onset and termination. The dynamic network as a whole is then represented as a collection of such spells.

To illustrate, with our hospital data hc, for each contact we might extend the contact period time index back by 20 s, indicating a tacit assumption that the corresponding contact lasted the entire 20-s period (in reality we know only that a contact occurred sometime during that period). Then for each pair of vertices in our network, a spell would be defined by the union of consecutive time periods, with the minimum length of any spell being 20 s, by definition.

```
#11.17 1 > hc.spls <- cbind((hc$Time-20)/(60*60),
       2 +                   hc$Time/(60*60),
       3 +                   hc$ID1,hc$ID2)
       4 > hc.dn <- networkDynamic(edge.spells=hc.spls)
```

Within **networkDynamic** it is then possible to query the network in various ways. For example, we can discover that the first edge (defined as the edge in our data with the earliest spell), while active during the first hour, was not active during the second hour.

---

[2]Using the option delete.vertices=TRUE in the function subgraph.edges would instead result in graphs with vertex subsets corresponding only to those people involved in at least one contact during a given 12-h period.

```
#11.18  1  > is.active(hc.dn,onset=0,terminus=1,e=c(1))
        2  [1] TRUE
        3  > is.active(hc.dn,onset=1,terminus=2,e=c(1))
        4  [1] FALSE
```

Similarly, for any given edge we can extract the full set of spells during which it was active. For instance, the first edge was active only once in fact, at the very start of our 4-day period.

```
#11.19  1  > get.edge.activity(hc.dn,e=c(1))
        2  [[1]]
        3                 [,1]           [,2]
        4  [1,]  0.03333333 0.03888889
```

On the other hand, the tenth edge was involved in seven contacts (i.e., seven spells of the minimal length of 20 s), all of which were confined to a roughly 24 h period.

```
#11.20   1  > get.edge.activity(hc.dn,e=c(10))
         2  [[1]]
         3                 [,1]           [,2]
         4  [1,]   0.800000   0.8055556
         5  [2,]   1.355556   1.3611111
         6  [3,]   1.505556   1.5111111
         7  [4,] 24.894444 24.9055556
         8  [5,] 25.005556 25.0166667
         9  [6,] 25.388889 25.3944444
        10  [7,] 25.500000 25.5055556
```

It is also possible within **networkDynamic** to convert between various different network-based representations. For example, analogous to earlier in this section, we can produce eight subnetworks corresponding to successive 12-h periods.

```
#11.21  1  > g.sl12.dN <- get.networks(hc.dn,start=0,end=96,
        2  +                                  time.increment=12)
```

Alternatively, we can summarize the network in the form of a data frame.

```
#11.22  1  > hc.dn.df <- as.data.frame(hc.dn)
        2  > names(hc.dn.df)
        3  [1] "onset"                "terminus"
        4  [3] "tail"                 "head"
        5  [5] "onset.censored"       "terminus.censored"
        6  [7] "duration"             "edge.id"
```

Here the rows correspond to spells, rather than contacts. We find that the majority of spells in fact lasted no more than one or at most two consecutive contact periods of 20 s. In contrast, the longest lasted for a bit over an hour.

```
#11.23  1  > summary(hc.dn.df$duration)
        2      Min.  1st Qu.   Median     Mean  3rd Qu.     Max.
        3  0.005556 0.005556 0.005556 0.012833 0.011111 1.088889
```

11.3   Visualization of Dynamic Networks

With the various possible dynamic network representations just discussed there come
a similar variety of ways in which dynamic networks may be visualized. Obviously
the vast majority of the basic principles of graph drawing or visualization that were
discussed in Chap. 3, for static network graphs, continue to hold true in the dynamic
setting. However, some additional concerns arise as well in the latter case, around
the question of how best to incorporate the notion of time.

Of course, we can ignore time altogether and simply visualize either a multi-graph
representation or the corresponding weighted graph obtained by collapsing multi-
edges. For the hospital data, plotting a multi-graph (not shown) with on the order of
32 thousand edges among just 75 vertices yields a visualization that is essentially
useless. A visualization of the corresponding weighted graph, shown in Fig. 11.2, is
somewhat better, but is still much too busy for us to gain any real insight.

```
#11.24  1 > detach(package:networkDynamic)
        2 > set.seed(42)
        3 > l = layout_with_fr(g.week.wgtd)
        4 > v.cols <- character(75)
        5 > v.cols[V(g.week.wgtd)$Status=="ADM"] <- "yellow"
        6 > v.cols[V(g.week.wgtd)$Status=="MED"] <- "blue"
        7 > v.cols[V(g.week.wgtd)$Status=="NUR"] <- "green"
        8 > v.cols[V(g.week.wgtd)$Status=="PAT"] <- "black"
        9 > plot(g.week.wgtd, layout=l, vertex.size=3,
       10 +     edge.width=2*(E(g.week.wgtd)$weight)/100,
       11 +     vertex.color=v.cols,vertex.label=NA,)
```

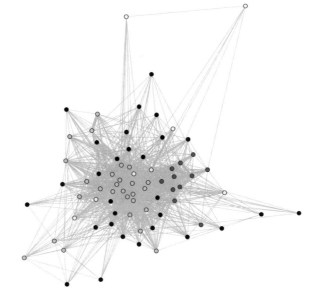

Fig. 11.2 Visualization of
the hospital contact data,
based on a weighted graph
summarizing all contacts
over the full week. Vertex
colors indicate status of each
person: ADM (*yellow*),
MED (*blue*), NUR (*green*),
and PAT (*black*)

Ideally, we would like to maintain as much of the available temporal information as possible when visualizing dynamic network data. From the point of view of the graph display problem, a fundamental challenge in moving from the static to the dynamic is the need to respect, in the case of the latter, what is referred to as the user's *mental map*. This term is used to describe the result of the process by which, upon studying a given static network map, a user becomes familiar with it, interprets it, and navigates about it. Simply put, we would expect a certain amount of 'stability' across visualizations.

When the dynamic network is represented as a sequence of 'slices', whether due to measurement constraints or user choice, and the number of slices is not too large, it may be useful to look at the corresponding sequence of static graph visualizations. In this case, it is natural to encourage a sense of stability by maintaining the positions of vertices across visualizations.

For example, in Fig. 11.3 we show plots of the hospital contact network as a sequence of eight static plots, each corresponding to a 12-h time period.

```
#11.25  1 > opar <- par()
        2 > par(mfrow=c(2,4),
        3 +       mar=c(0.5, 0.5, 0.5, 0.5),
        4 +       oma=c(0.5, 1.0, 0.5, 0))
        5 > for(i in (1:8)) {
        6 +   plot(g.sl12[[i]], layout=l, vertex.size=5,
        7 +        edge.width=2*(E(g.week.wgtd)$weight)/1000,
        8 +        vertex.color=v.cols,vertex.label=NA)
        9 +   title(paste(12*(i-1),"to",12*i,"hrs"))
       10 + }
       11 > par(opar)
```

More ambitiously, if the data are rich enough temporally, instead of individually displaying each of the many corresponding individual static graphs, one might animate them to produce a 'movie,' so as to watch the network graphs change over time. Although the area of dynamic graph drawing of this type is arguably still young, there are already a number of approaches that have been developed for automated display of dynamic network graphs with constraints to reinforce a consistent user mental map. Broadly speaking, rather than stringently refusing to allow vertices and edges to move, as above, one attempts to allow only 'local' elements in the vicinity of a newly added graph element to change.

Within R, at the time of this writing the primary package that allows for the production of such dynamic network visualizations is **ndtv**, which takes as its input an object from **networkDynamic**. However, it should be noted that the process of creating a dynamic layout can take considerably more time than in the static case, and the authors of **ndtv** caution that its use currently is most realistic for networks of relatively small size. For example, it is unrealistic to apply to our hospital contact data.

On the other hand, it is possible to produce a useful visualization of primarily the temporal information in dynamic network data in a relatively straightforward

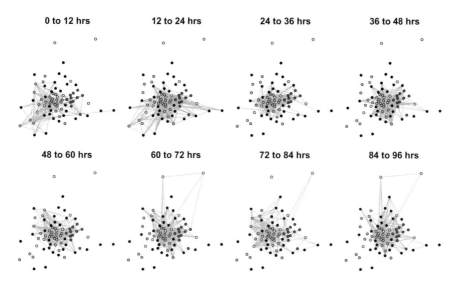

Fig. 11.3 Visualization of the hospital contact data, as a sequence of eight static networks corresponding to consecutive 12-h periods. Vertex colors indicate status of each person: ADM (*yellow*), MED (*blue*), NUR (*green*), and PAT (*black*)

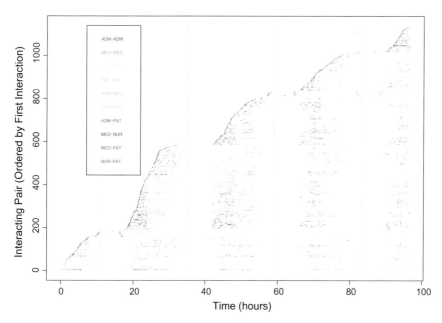

Fig. 11.4 Timeline showing the presence/absence of dynamic edges in the hospital contact network, covering just over 96 h over 5 days (midnight of each day indicated by *gray vertical lines*)

manner, if connectivity beyond vertex pairs (i.e., edges) is suppressed. Specifically, we can display the onset and termination of edges as a function of time, in the form of a type of phase-plot, with the edges indexed along the y-axis in order of their first appearance in the network. A visualization of this type is shown in Fig. 11.4, for the hospital contact data.[3]

```
#11.26  1  > # Establish colors for edge status.
        2  > tmp.es <- paste(v.status[hc.dn.df$tail],"-",
        3  +                  v.status[hc.dn.df$head],sep="")
        4  > mycols <- numeric(nrow(hc.dn.df))
        5  > mycols[tmp.es=="ADM-ADM"] <- 1
        6  > mycols[tmp.es=="MED-MED"] <- 2
        7  > mycols[tmp.es=="NUR-NUR"] <- 3
        8  > mycols[tmp.es=="PAT-PAT"] <- 4
        9  > mycols[(tmp.es=="ADM-MED") | (tmp.es=="MED-ADM")] <- 5
       10  > mycols[(tmp.es=="ADM-NUR") | (tmp.es=="NUR-ADM")] <- 6
       11  > mycols[(tmp.es=="ADM-PAT") | (tmp.es=="PAT-ADM")] <- 7
       12  > mycols[(tmp.es=="MED-NUR") | (tmp.es=="NUR-MED")] <- 8
       13  > mycols[(tmp.es=="MED-PAT") | (tmp.es=="PAT-MED")] <- 9
       14  > mycols[(tmp.es=="NUR-PAT") | (tmp.es=="PAT-NUR")] <- 10
       15  > my.palette <- rainbow(10)
       16  > # Produce plot.
       17  > ne <- max(hc.dn.df$edge.id)
       18  > max.t <- max(hc.dn.df$terminus)
       19  > plot(c(0,max.t),c(0,ne),ann=F,type='n')
       20  > segments(hc.dn.df$onset,hc.dn.df$edge.id,
       21  +   hc.dn.df$terminus,hc.dn.df$edge.id,
       22  +   col=my.palette[mycols])
       23  > title(xlab="Time (hours)",
       24  +   ylab="Interacting Pair
       25  +   (Ordered by First Interaction)")
       26  > clip(0,max.t,0,ne)
       27  > abline(v=c(11,35,59,83),lty="dashed",lw=2,
       28  +   col="lightgray")
       29  > # Add legend to plot.
       30  > status.pairs <- c("ADM-ADM","MED-MED","NUR-NUR",
       31  +   "PAT-PAT", "ADM-MED","ADM-NUR","ADM-PAT",
       32  +    "MED-NUR", "MED-PAT","NUR-PAT")
       33  > legend(7,1140,status.pairs,
       34  +         text.col=my.palette[(1:10)],cex=0.7)
```

Some of the temporal information in Fig. 11.1 is evident as well here in Fig. 11.4. For example, the diurnal patterns in contacts are clear. However, here we now have the capability to track individual edges and, through the use of color, edges of various combinations of statuses. For example, we see that there is a cluster of contacts between nurses that are introduced in the second half of the 2 day (i.e., edges numbered roughly 500, just after hour twenty) that are not seen again until late in the last day.

---

[3]The function `timeline` in **ndtv** can be used to produce a similar plot, but currently does not have the functionality to permit annotation by edge attributes, as we require here.

11.4 Characterization of Dynamic Networks

The development of a comprehensive body of tools for characterizing dynamic networks, and of a corresponding understanding of their properties and usage, lags far behind that for static graphs G, such as was described in Chap. 4. This lacking is probably due in part to insufficiently pressing need, in that traditionally it has been notably more difficult in most network studies to obtain time-indexed network data that are rich in both quality and quantity. As we have noted, however, this situation is changing, and the demand for tools to characterize dynamic networks is growing. Also contributing to the gap between static and dynamic methods of characterization is possibly the absence of a similarly mature graph-theoretic infrastructure. Finally, another important factor is the sheer combinatorial explosion in problem variants, once the factor of 'time' is to be incorporated. See Moody et al. [4], for example, for a detailed discussion of how and why such variants arise, in the context of social networks.

At present, therefore, arguably the most common manner in which dynamic networks are characterized in practice is through the application of methods for static networks to successive slices over time. Just as in the static case, the choice of what methods to use is specific to the application at hand and should be driven by particular

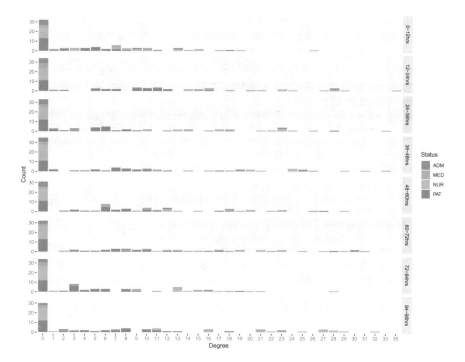

Fig. 11.5 Degree distributions for the hospital contact data, one per each of eight static networks corresponding to consecutive 12-h periods

questions about the complex system being studied. Here we illustrate this approach
to dynamic network characterization using the hospital contact data again, where
recall that the goal was to obtain insight into the manner in which infection might be
transmitted between individuals in this environment.

In Sect. 8.5.1 we saw that network topology can affect the spread of a disease. And
we know from Chap. 4 that the degree distribution is one of the most fundamental
summaries of network topology. Thus, to begin the process of understanding the
potential spread of infection in the hospital from which these data were obtained, we
examine the degree distributions for each of the eight successive 12-h periods.

```
#11.27  1 > all.deg <- sapply(g.sl12,degree)
        2 > sl.lab<- sapply(1:8, function(i)
        3 +    paste(12*(i-1), "-", 12*i, "hrs", sep=""))
        4 > deg.df <- data.frame(Degree=as.vector(all.deg),
        5 +    Slice = rep(sl.lab,each=75),
        6 +    Status = rep(V(g.week)$Status, times=8))
        7 > library(ggplot2)
        8 > p = qplot(factor(Degree), data=deg.df,
        9 +                geom="bar",fill=Status)
       10 > p+facet_grid(Slice~.) + xlab("Degree") + ylab("Count")
```

In Fig. 11.5 are shown the resulting displays of the degree distributions, visualized
as bar charts, where the bars have been colored to indicate the relative frequency of
vertices of each of the four statuses. Several interesting features are evident. We
see, for example, a large mode at zero in each bar chart. During any 12-h period
there is a substantial fraction of the 75 people studied that were not involved in
any recorded contacts. Among hospital employees (i.e., ADM, MED, and NUR)
this pattern presumably reflects duty shifts. More interestingly, we see that the vast
majority of patients (i.e., PAT) are involved in contacts with comparatively fewer
numbers of other people. Similarly, the hospital employees tend to have contacts
with higher numbers of people.

This exploration suggests that the most pervasive potential spreaders would be
among the hospital employees. In order to try to identify such individuals, and
acknowledging that they may change from day to day, we create a summary of
who is consistently among the top five (i.e., has the fifth highest degree or larger)
over the week.

```
#11.28  1 > top.deg <- lapply(1:8,function(i) {
        2 +    all.deg[,i][rank(all.deg[,i])>=70]
        3 + })
        4 > table(unlist(lapply(1:8,function(i)
        5 +    as.numeric(names(top.deg[[i]])))))
        6   1  5  7  8 10 11 13 15 17 19 21 22 23 24 25 26 27 29 31
        7   2  2  4  1  1  2  2  4  3  3  2  1  3  2  1  2  2  2  1
        8  34 36 37 63 64
        9   1  1  2  1  2
```

We see that two of the people in this network (i.e., with vertex IDs 7 and 15) are among
the top five during four of the eight successive 12-h periods. And these two are indeed
among the hospital employees, consisting of a nurse and a doctor, respectively.

```
#11.29  1 > V(g.week)$Status[c(7,15)]
        2 [1] "NUR" "MED"
```

Next, for these two individuals, we look at the relative number of contacts per person contacted (i.e., the number of contacts per alter, for each of these two egos). This means looking at ratio of vertex strength to degree.

```
#11.30 1 > all.str <- sapply(g.sl12,strength)
       2 > all.r <- all.str/all.deg
       3 > round(all.r[c(7,15),],digits=2)
       4      [,1]  [,2] [,3]  [,4]  [,5]  [,6]  [,7]  [,8]
       5 7   2.00 25.79 11.1 32.73 14.20 33.19  8.33 37.34
       6 15 29.71 26.33 17.0 12.48 19.27 15.30 19.40 12.93
```

We see that, compared to the network as a whole,

```
#11.31 1 > summary(c(all.r))
       2    Min. 1st Qu.  Median   Mean 3rd Qu.    Max.    NA's
       3    1.00    5.00   10.43  12.06   15.73   47.35     261
```

this nurse and this doctor tend largely to have a rate of contacts at or above the mean, and more often than not, in the upper quartile. So not only do these two individuals come into contact with many more others than usual in this hospital ward, they also tend to have noticeably higher rates of contact with each, on average.

Accordingly, one might argue that these two individuals are natural candidates to consider as potential 'super-spreaders'. It would then be natural to also consider just how quickly an infection might spread from these two. One rough indicator is the average shortest-path distance from each of these two to the rest of the individuals in the network. Again calculating this quantity slice-by-slice,

```
#11.32  1 > sp.len <- lapply(1:8, function(i) {
        2 +    spl <- distances(g.sl12[[i]],v=c(7,15),
        3 +                        to=V(g.sl12[[i]]),
        4 +                        weights=NA)
        5 +    spl[spl==Inf] <- NA
        6 +    spl
        7 + })
        8 > ave.spl <- sapply(1:8,function(i)
        9 +    apply(sp.len[[i]],1,mean,na.rm=T))
       10 > round(ave.spl,digits=2)
       11     [,1] [,2] [,3] [,4] [,5] [,6] [,7] [,8]
       12 7  3.05 1.27 1.79 1.12 1.80 1.33 2.00 1.24
       13 15 1.72 1.51 2.48 1.35 1.36 1.26 1.61 1.36
```

we find that for both individuals in most slices their distance on average from the rest of the network—among those having at least one contact during that slice—is less than two. These numbers might seem quite small compared to the overall diameter of the networks.

```
#11.33 1 > sapply(g.sl12,diameter)
       2 [1]  9  8 26 28 10 10 10 10
```

However, compared to the overall average shortest path distance, the contrast is decidedly less starck.

```
#11.34 1 > round(sapply(g.sl12,mean_distance),digits=2)
       2 [1] 2.12 1.70 1.81 1.67 1.79 1.70 1.91 1.78
```

Nevertheless, both the nurse and the doctor do still seem to be somewhat closer to the others in the network most of the time, which could be argued to likely have a further detrimental effect on containing infection, were either of them to become infected themselves.

11.5   Modeling Dynamic Networks

The modeling of dynamic network data is nontrivial. The complexity that we have already seen associated with modeling a static network graph, in Chaps. 5 and 6, is further magnified by the need to not only model 'many' such graphs (i.e., in a discrete or a continuous sense, as the case may be), but also—in particular—by the need to model at the same time the manner in which these graphs evolve from one to another. Compared to the existing literature on modeling of static network graphs, there is decidedly less development on the dynamic side. But this state of affairs is changing. However, as of this writing, there is little available in R. And, in addition, a sufficiently careful treatment of the few dynamic network modeling frameworks that are available in R would require substantial attention and is beyond the scope of this book. Therefore, we will restrict ourselves here simply to summarizing two available options.

Given the important role that exponential random graph models (ERGMs) have played in the modeling of static network graphs, as described in Chap. 6, they have been considered a natural candidate for extension to the case of dynamic networks. Building upon and modifying earlier work of Hanneke and Xing [1] and Hanneke et al. [2], Krivitsky and Handcock [3] introduced a class of temporal exponential random graph models for longitudinal network data (i.e., in which networks are observed in panels, like the 'slices' we created for the hospital data), in which the formation and dissolution of edges are modeled in a separable fashion.

Specifically, let $\mathbf{Y}(t)$ and $\mathbf{Y}(t-1)$ denote the adjacency matrices for the network at times t and $t-1$, and (in a slight abuse of notation) define the so-called formation network $\mathbf{Y}^+ = \mathbf{Y}(t) \cup \mathbf{Y}(t-1)$ and dissolution network $\mathbf{Y}^- = \mathbf{Y}(t) \cap \mathbf{Y}(t-1)$ relative to time point t. That is, $\mathbf{Y}^+$ is the initial network $\mathbf{Y}(t-1)$ plus the additional ties formed by time t, and $\mathbf{Y}^-$ is the initial network $\mathbf{Y}(t-1)$ less the ties dissolved by time t. The separable, temporal ERGM (i.e., STERGM) framework of Krivitsky and Handcock then assumes (a) a discrete-time Markov structure for the evolution of the network over time, (b) that the formation and dissolution networks are conditionally independent (i.e., separability), and (c) an exponential family form for the relevant conditional distributions. The result is a model of the form

$$\mathbb{P}_\theta \left(\mathbf{Y}(t) = \mathbf{y}(t) \mid \mathbf{Y}(t-1) = \mathbf{y}(t-1) \right) = \mathbb{P}_\theta \left(\mathbf{Y}^+ = \mathbf{y}^+ \mid \mathbf{Y}(t-1) = \mathbf{y}(t-1) \right)$$
$$\times \, \mathbb{P}_\theta \left(\mathbf{Y}^- = \mathbf{y}^- \mid \mathbf{Y}(t-1) = \mathbf{y}(t-1) \right) \, ,$$

where the two conditional distributions on the right-hand side above have exponential family form (i.e., as in Eq. 6.1).

The specification of effects for these models can, like the ERGMs of Chap. 6, involve both network effects (i.e., endogenous) and effects due to vertex or edge attributes (i.e., exogenous). Maximum likelihood estimates of the corresponding parameters θ may then be calculated using extensions of the methods for standard ERGMs, based on Monte Carlo. The R package **tergm** contains a collection of functions that allow for simulation, fitting, and diagnostics with this class of models.

Alternatively, one might think to replace the discrete-time formulation in temporal ERGMs with a continuous-time formulation. The stochastic actor-oriented dynamic networks models of Snijders and colleagues do just that. See [5], for example, for an overview. Under these models, a continuous-time Markov chain is formulated, with each possible network corresponding to a state. The rate functions governing waiting times between state transitions (i.e., from one network to another, where the two differ in the presence or absence of a single edge) are expressed as the sum of the collection of edge-specific rates. Each of the latter are in turn represented in the manner of a generalized linear model (GLM), allowing for the incorporation of specific edge and/or vertex effects. Again, a Monte Carlo approach is required for simulation, fitting, and diagnostics. Techniques for these tasks have been developed over a period of years and are now available in the R package **RSiena**.

References

1. S. Hanneke and E. P. Xing, "Discrete temporal models of social networks," in *Statistical Network Analysis: Models, Issues, and New Directions.* Springer, 2007, pp. 115–125.
2. S. Hanneke, W. Fu, and E. P. Xing, "Discrete temporal models of social networks," *Electronic Journal of Statistics*, vol. 4, pp. 585–605, 2010.
3. P. N. Krivitsky and M. S. Handcock, "A separable model for dynamic networks," *Journal of the Royal Statistical Society: Series B (Statistical Methodology)*, 2013.
4. J. Moody, D. McFarland, and S. Bender-deMoll, "Dynamic network visualization," *American Journal of Sociology*, vol. 110, no. 4, pp. 1206–1241, 2005.
5. T. A. Snijders, G. G. Van de Bunt, and C. E. Steglich, "Introduction to stochastic actor-based models for network dynamics," *Social networks*, vol. 32, no. 1, pp. 44–60, 2010.
6. P. Vanhems, A. Barrat, C. Cattuto, J.-F. Pinton, N. Khanafer, C. Régis, B.-a. Kim, B. Comte, and N. Voirin, "Estimating potential infection transmission routes in hospital wards using wearable proximity sensors," *PloS one*, vol. 8, no. 9, p. e73970, 2013.

Index

© Springer Nature Switzerland AG 2020
E. D. Kolaczyk and G. Csárdi, *Statistical Analysis of Network Data with R*, Use R!,
https://doi.org/10.1007/978-3-030-44129-6

Printed in the United States
By Bookmasters